T0212857

Jesuits and the Natural Sciences in Modern Times, 1814–2014

Brill Research Perspectives in Jesuit Studies

Volumes published in this Brill Research Perspectives title are listed at *brill.com/rpjs*

Jesuits and the Natural Sciences in Modern Times, 1814–2014

By

Agustín Udías

BRILL

LEIDEN | BOSTON

This paperback book edition is simultaneously published as issue 1.3 (2019) of *Brill Research Perspectives in Jesuit Studies*, DOI:10.1163/25897454-12340003.

Library of Congress Control Number: 2019939146

Typeface for the Latin, Greek, and Cyrillic scripts: "Brill". See and download: brill.com/brill-typeface.

ISBN 978-90-04-39489-6 (paperback)
ISBN 978-90-04-39490-2 (e-book)

This book is printed on acid-free paper and produced in a sustainable manner.

Contents

Jesuits and the Natural Sciences in Modern Times, 1814–2014   1
   Agustín Udías
   Abstract   1
   Keywords   1
1   Introduction   1
2   A New Beginning   3
3   Science in the Training of Jesuits and the Tension between
       Scholastic Philosophy and Modern Science   5
       3.1   *Evolution of Teaching Science to Jesuit Students*   8
4   Science in Jesuit Universities, Colleges, and Secondary Schools   12
       4.1   *Jesuits Teaching Sciences in Jesuit Universities*   13
       4.2   *Science Teaching in Secondary Schools*   15
       4.3   *Science Textbooks*   16
5   The New Observatories   17
       5.1   *Observatories in Europe and United States*   22
       5.2   *Observatories in Asia, Africa, Australia, and Central and
             South America*   29
       5.3   *The Vatican Observatory*   38
6   The Earth's Magnetism   40
7   Jesuit Meteorological Stations   43
8   Tropical Hurricanes   49
       8.1   *Benito Viñes and Caribbean Hurricanes*   49
       8.2   *Typhoons and Cyclones in the Philippines and South China Sea*   52
9   Earthquakes and Seismology   59
       9.1   *Seismographic Stations in Europe*   59
       9.2   *The Jesuit Seismological Association and Jesuit Seismologists in
             North America*   62
       9.3   *Jesuit Seismologists in Central and South America*   68
       9.4   *Seismological Stations in Asia, Africa, and Oceania*   72
       9.5   *The Study of Microseisms and Seismological Organizations*   73
10   The Tradition in Mathematics   76
11   The New Naturalists and Biologists   77
       11.1 *Botany and Ecology in India*   81
12   Jesuit Scientists in Non-Jesuit Institutions   83
13   Recent Developments   86
14   Jesuit Scientists and Ignatian Spirituality   91
15   Conclusion   96
Bibliography   97

Jesuits and the Natural Sciences in Modern Times, 1814–2014

Agustín Udías
Universidad Complutense de Madrid, Spain
agustinudias@gmail.com

Abstract

After their restoration of 1814, the Jesuits made significant contributions to the natural sciences, especially in the fields of astronomy, meteorology, seismology, terrestrial magnetism, mathematics, and biology. This narrative provides a history of the Jesuit institutions in which these discoveries were made, many of which were established in countries that previously had no scientific institutions whatsoever, thus generating a scientific and educational legacy that endures to this day. The essay also focuses on the teaching and research that took place at Jesuit universities and secondary schools, as well as the order's creation of a worldwide network of seventy-four astronomical and geophysical observatories where particularly important contributions were made to the fields of terrestrial magnetism, microseisms, tropical hurricanes, and botany.

Keywords

Jesuit scientists – Jesuit observatories – Jesuit universities – seismology – astronomy – meteorology – hurricanes – biology – Ignatian spirituality

1 Introduction

Pope Clement XIV's (1705–74, r.1769–74) suppression of the Society of Jesus in 1773, which followed the Jesuits' suppression in Portugal (1758), France (1764), and Spain and its American colonies (1767), put an end to the scientific work in which the order had been involved for more than two hundred years. All of a sudden, the work of numerous Jesuit scientists in European, Asian, and

American institutions was cut short.[1] Tragically, the order's suppression also coincided with the Jesuits' involvement in the science of astronomy, with the suppression being implemented at a time when the order was just beginning to open astronomical observatories in its colleges and universities and in a period when it had begun introducing Newtonian physics and astronomy into the Jesuit curriculum. The suppression also brought an end to the Jesuits' efforts to introduce European scientific knowledge, especially astronomy, to China and India, as well as their pioneering work of presenting the flora, fauna, and geography of the newly discovered countries of America to a European audience for the very first time. Although some Jesuits were able to continue their work after the suppression, whether privately or in state universities and observatories, the rarity of these stories of survival ultimately confirm that this was indeed the end of an age.

However, in 1814, Pope Pius VII (1742–1823, r.1800–23) restored the Society of Jesus, inaugurating a new chapter in the history of the Jesuit contribution to the natural sciences. In the period that followed, the Jesuits made important contributions to the fields of astronomy, meteorology, geomagnetism, seismology, mathematics, and biology through their research at newly established Jesuit observatories, colleges, and universities.

Although this work continues to the present day,[2] the Jesuits' involvement in science after 1814 has elicited far less interest from historians than the pre-suppression period. The few publications to have examined Jesuit scientific work after the restoration include the work of the current author, *Jesuit Contribution to Science: A History* (2014);[3] George Bishop's brief history of Jesuit involvement in science and mathematics;[4] and the work of Jaime Fernández and Eduardo di Vita,[5] which is actually a list of Jesuit scientists with information reduced to succinct biographies and a short list of publications.

---

1   References for this period can be found in Sheila J. Rabin, "Jesuit Science before 1773: A Historiographical Essay," *Jesuit Historiography Online*, ed. Robert A. Maryks, https:// referenceworks.brillonline.com/entries/jesuit-historiography-online/jesuit-science-before-1773-a-historiographical-essay-COM_196375 (accessed August 2, 2018).

2   Agustín Udías, "Jesuit Contribution to Science 1814–2000: A Historiographical Essay," *Jesuit Historiography Online*, ed. Robert A. Maryks; http://referenceworks.brillonline. com/entries/jesuit-historiography-online/jesuit-contribution-to-science-18142000-a-historiographical-essay-COM_192552 (accessed August 2, 2018).

3   Agustín Udías, *Jesuit Contribution to Science: A History* (Dordrecht: Springer, 2015), and *Los jesuitas y la ciencia: Una tradición en la Iglesia* (Bilbao: Mensajero, 2014).

4   George D. Bishop, *Jesuit Pioneers of Modern Science and Mathematics* (Anand, Gujarat: Gujarat Sahitya Prakash, 2005).

5   Jaime Fernández Rodríguez and Eduardo di Vita, *Presença dos jesuítas no mundo científico* (São Paulo: Edições Loyola, 2004).

Building on my previous work for the period 1814–2014, this paper provides a summary of Jesuit contributions to the natural sciences after the restoration, especially in the fields of astronomy, meteorology, seismology, terrestrial magnetism, mathematics, and biology. The essay also provides a history of the vast network of Jesuit institutions in which these discoveries were made, many of which were established in countries that previously had no scientific institutions whatsoever, thus generating a scientific and educational legacy that endures to this day.

In detailing the Jesuits' contributions to science during this period, the essay also highlights the exceptional nature of the order itself insofar as its members were often at the forefront of the latest scientific discoveries, which differentiates the Society from other religious orders. Thus, one of the questions the paper seeks to address is why the Society of Jesus, as a religious order, attached so much importance to its members engaging in scientific research. To a certain extent, the answer to this question can be found in the Jesuits' desire to return to the scientific tradition of the sixteenth and eighteenth centuries, when the order established chairs of mathematics in their colleges and universities, thus contributing to the beginning of modern science. At the same time, the Jesuits' involvement in science in the post-restoration period also served an apologetic aim in countering the secular ideologies of the modern age. However, though numerous Jesuit scientists did indeed produce works seeking to demonstrate that there was no opposition between science and faith, the paper argues that it is impossible to understand the order's support for scientific research without taking into account the "Ignatian spirituality" that is at the heart of every Jesuit's spiritual life. While the extent of the order's encouragement of scientific research and the involvement of Jesuits in scientific work has ebbed and flowed over time, the essay argues that is this spirituality that has ultimately been the driving force of the Jesuit presence in science, both in the modern and early modern periods.

2   A New Beginning

During the almost fifty years in which the Society was suppressed (1773–1814), important progress had been made in practically all fields of the natural sciences, and profound changes had also taken place in society more generally. These new developments conditioned the Jesuits' scientific work following the restoration, lending it a different character from that of the previous period (1540–1773). It was a new beginning.

In the nineteenth century, scientific and technological progress had laid the foundations for the industrial revolution that would transform modern society. In almost all fields of science, major developments had taken place in both the theory and practice of science. As a result, the Jesuit scientists of the restored Society encountered a form of science that differed from that of the previous century, with large numbers of scientists working in already established fields. This may explain why Jesuit research was more often to be found in sciences such as meteorology, seismology, and biology, which were then being developed, rather than in physics and chemistry, which were already established.

Radical changes had also taken place in education, a sphere in which the Jesuits had previously excelled, having founded a vast network of colleges before the suppression. Medieval universities in Europe, which had suffered a general decline during the eighteenth century, were in a profound process of reform, beginning in Germany and France. As part of this process, most universities now attached an increasing amount of importance to scientific research. And the nature of education itself had also changed radically, as education was increasingly viewed as a public service to be delivered by the state, the provision of which was often underpinned by a secular ethos. Thus governments established new state institutions with official regulations and study programs with a definite emphasis on the natural sciences that were to be followed at all three levels (i.e., primary, secondary, and tertiary or university level). Having lost all of their European schools and universities (nearly six hundred in total), the Jesuits needed to adapt their schools to these programs and this entirely new context, in the process abandoning, at least in part, the humanist emphasis that had underpinned their colleges and schools before the suppression.

Due to these changes, the traditional Jesuit program, following the norms of the *Ratio studiorum* (the plan of studies used at Jesuit colleges, the definitive version of which was published in 1599), with its three stages of humanities, philosophy, and theology, was no longer suitable for modern educational institutions. While an attempt was made to revise the *Ratio* by placing a greater emphasis on the teaching of mathematics and natural sciences, this soon proved a futile task, as educational programs at state universities and secondary schools varied greatly from one country to the next, making it impossible to impose a single model. Hence the need to adapt Jesuit schools to official educational norms, with different types of programs in different countries, resulted in the old *Ratio* losing its relevance.

In the meantime, and partly as a result of these changes in education and science, the Catholic Church increasingly found itself under attack from those who claimed that science and Catholicism were fundamentally opposed to one another. Accordingly, the Jesuits' involvement in scientific research often had an apologetic aim in countering the idea that science and faith were incompatible, an idea that had become particularly widespread from the mid-nineteenth century onward. In 1923, for example, Aloysius Cortie (1859–1925), the director of the Stonyhurst Observatory (Lancashire, England), referred to the work of the Italian Jesuit astronomer Angelo Secchi (1818–78) as a clear example that religion and science were not mutually exclusive:

> The enemies of the Holy Church have made such unwarranted use of science as a weapon of attack against her most fundamental truths that an impression has sometimes been introduced among many of her children that the pursuit of science is damaging and dangerous to faith [...]; Father Secchi is a striking example of one who knew how to unite religion and science.[6]

3   Science in the Training of Jesuits and the Tension between Scholastic Philosophy and Modern Science

In order to meet the demands of these changes in science and education, the Jesuits would clearly need to receive a thorough scientific education themselves. Unfortunately, in the post-1814 period, and in contrast to the situation in the Jesuit provision of education more broadly, the training of Jesuits largely followed the pre-suppression system and hence differed from the education provided to students in the new colleges and universities. In the Old Society, scientific studies, including mathematics, astronomy, and physics, were part of the philosophy program, the second stage in Jesuit training before the study of theology. In the eighteenth century, some modern topics in advanced mathematics and Newtonian physics and astronomy had already been introduced as part of the courses on natural philosophy and mathematics. Yet there had been substantial changes and developments in these sciences in the intervening period. Initially, the order sought to respond to these developments by incorporating lectures on mathematics, astronomy, physics, chemistry, biology, and geology into Jesuit training. However, this in turn fed into a broader controversy surrounding the role of Scholastic philosophy in the Jesuits' training.

In the initial years following the restoration, there was a debate at the Roman College between those who argued that modern science should be

---

6   Aloysius L. Cortie, "Angelo Secchi, S.J. (1818–1878)," in *Twelve Catholic Men of Science*, ed. Bertram Windle (London: Catholic Truth Society, 1923), 149–64.

taught separately from philosophy and those who proposed returning to the teaching of Aristotelian natural philosophy on the basis of neo-Scholasticism or the neo-Thomist school.[7] Giovanni Battista Pianciani (1784–1862), professor of physics and chemistry between 1824 and 1848, and his successor Francesco Saverio Provenzali (1815–91), were prominent defenders of the first position. Secchi, who was director of the observatory and taught astronomy at the college, gave them his backing. Pianciani, the author of a textbook entitled *Istituzioni fisico-chimiche* (Physico-chemical principles, 4 vols. [1833–35]), insisted on the need for teaching in modern physics and chemistry. Two other professors at the Roman College, Salvatore Tongiorgi (1820–65) and Domenico Palmieri (1829–1909), also supported this position and rejected Aristotelian physics, especially the theory of hylomorphism defended by the neo-Thomist philosophers.

Matteo Liberatore (1810–92), author of *Institutiones philosophicae* (Philosophical principles [1860–61])—a highly influential work published in several editions—and Giovanni Cornoldi (1822–92), who strongly criticized Secchi's book *L'unità delle forze fisiche* (The unity of physical forces [1864]) on the basis that it was contrary to Aristotelian doctrine, were among the defenders of the neo-Thomist doctrine. One of the key issues in this controversy was the Scholastic theory of the hylomorphic (matter–form) composition of material bodies as opposed to the theory of atomic composition, supported by the discoveries of modern physics and chemistry. Secchi, who defended the atomic theory, complained that the ideas of Roger Boscovich (1711–87), a professor at the Roman College in the period before the Society's suppression, had found little acceptance among Jesuit professors of natural philosophy. Boscovich, who accepted the claims of Newtonian physics, had presented a dynamic atomic theory where atoms are centers of forces—which was consistent with modern atomism and opposed to Scholastic natural philosophy—in his book *Theoria philosophiae naturalis* (Theory of natural philosophy [1758]). Liberatore, on the other hand, defended Scholastic philosophy and hylomorphism and strongly attacked Tongiorgi for accepting the atomism of modern chemistry.

At first, the order responded to this debate by calling for moderation on both sides and for Scholasticism and the natural sciences to be accorded equal weight in the Jesuits' training. In the words of decree 17 of the Twenty-First General Congregation (1829): "Care must be taken that the study of mathematics and physics should not be in detriment of that of philosophy [i.e.,

---

7   Robert Jacquin, "Une polémique romaine sur l'hylémorphisme il-y-a un siècle," *Revue des sciences religieuses* 32 (1958): 372–77.

Scholasticism]."[8] A further intervention was then made in 1863, when Superior General Peter Beckx (1795–1887, in office 1853–87) issued a conciliatory letter requesting that freedom of opinion be respected in cases where there was doubt. However, Beckx later changed his mind and in 1878 took the side of Liberatore in calling for a greater emphasis on Scholastic natural philosophy in Jesuit training. This decision eventually received papal support in the encyclical *Aeterni Patris* (1879) of Pope Leo XIII (1810–1903, r.1878–1903), which defended neo-Thomism and demanded that Christian philosophy be taught on the basis of the doctrine of St. Thomas Aquinas (c.1225–74), claiming that it was a necessary preparation for theology. Yet it is important to note that the pope was not opposed to teaching science per se; indeed, the encyclical's conclusion states that

> nothing is of greater use to the philosopher than diligently to search into the mysteries of nature and to be earnest and constant in the study of physical things [...]; between certain and accepted conclusions of modern physics and the philosophical principles of the schools there is no conflict worthy of the name.[9]

In their Twenty-Third General Congregation (1883), the Jesuits complied with the encyclical and stated that nothing should be explained in experimental physics that contradicted the doctrine of St. Thomas (decree 15), although the teaching of natural and physical sciences was to continue (decree 17).[10] Both approaches had to be respected in the study of nature: the philosophical, in accordance with neo-Scholastic Thomism, and the scientific, in accordance with the modern approach of the positive sciences.

During this period, the part of natural philosophy that deals with the nature of material objects, required by the above-mentioned encyclical, began to be known as "cosmology" (not to be confused with the modern use of this term). Thus, cosmology became a part of the Scholastic theses of neo-Thomism, which were adopted in the Jesuit teaching of philosophy. Cosmology was viewed as a different approach to the knowledge of natural phenomena from the modern natural sciences. This was consequently the basis on which cosmology was taught at Jesuit faculties of philosophy, and it also informed the

8   *Institutum Societatis Iesu*, vol. 2, *Examen et constitutiones: Decreta congregationum generalium; Formulae congregationum* (Florence: Typographia a SS. Conceptione, 1893), 480.

9   http://w2.vatican.va/content/leo-xiii/en/encyclicals/documents/hf_l-xiii_enc_04081879_aeterni-patris.html (accessed September 3, 2018).

10   *Institutum Societatis Iesu*, 2:501–4.

work of individual Jesuit authors. An illustrative example is the popular text-book *Institutiones philosophicae* (Philosophical principles; 8 vols. [1890–1900]) by Juan José Urráburu (1844–1904), a professor at the Gregorian University, which contains a section dedicated to Scholastic cosmology that combines the philosophy of Francisco Suárez (1548–1617), who defended a liberal interpretation of Aristotelianism, with neo-Thomism. Another example is the popular Jesuit textbook of neo-Thomist cosmology entitled *Philosophia naturalis* (Natural philosophy [1894]) by Heinrich Haan (1844–1909), who taught in the German Faculty of Philosophy at Valkenburg in the Netherlands. A clear separation was thereby established between so-called Scholastic cosmology and the modern sciences of physics, chemistry, and biology.

3.1   *Evolution of Teaching Science to Jesuit Students*

As a result of this debate, science continued to be taught as part of the Jesuits' training in the philosophy program; however, its importance gradually diminished over time before being removed entirely from the philosophy program in the early twentieth century. Between 1850 and 1873, for example, there were still courses at the Roman College in mathematics, mathematical physics, physical chemistry, and astronomy. After the Italian government's confiscation of the Roman College in 1873, the Jesuit faculties of philosophy and theology became the Gregorian University, named in honor of the first patron, Gregory XIII (1502–85, r.1572–85), and science continued to be taught there until around the 1930s, when it came to an end after the publication of the regulation of ecclesiastic studies in 1931, as we will see below. In the early years of the twentieth century, among the last professors of science at that time, mention should be made of Giuseppe Gianfranceschi (1875–1934), who had a doctorate in physics and mathematics from the University of Rome and carried out research in electro-physics and acoustics. His book, *La fisica dei corpuscoli* (Physics of corpuscles [1916]), dealing with atomic physics, was highly popular, appearing in three editions and a Spanish translation. Gianfranceschi was the first director of Vatican Radio and in 1928 took part in Umberto Nobile's (1885–1978) airship expedition to the North Pole.

Science teaching also continued in other Jesuit faculties of philosophy during this period. One of them was the German philosophy and theology faculty at Valkenburg, where science was taught to a high standard from 1872 to 1941, especially in physics. One of the most prominent professors at Valkenburg was Theodor Wulf (1868–1946), who was the author of several physics textbooks as well as a book on Albert Einstein's (1879–1955) theory of relativity (*Einstein Relativitäts Theorie* [Einstein's theory of relativity (1912)]). His physics textbook (*Lehrbuch der Physik* [Textbook of physics (1926)]) went into several editions and was translated into English, French, and Spanish. As part of his research at Valkenburg, Wulf built a sensitive electrometer, capable of detecting very low levels of radiation,[11] which he used to carry out an experiment on the Eiffel Tower in 1910 that led to the discovery that radiation was stronger at the top than at the bottom, contrary to what had been expected. As a result of the experiment, Wulf concluded that radiation came not from the earth but from outside the atmosphere, thus becoming one of the first to discover the effect of cosmic rays, and he is now considered a pioneer in their study.[12] Another prominent physics professor at Valkenburg was Ludwig Dressel (1840–1917), author of a popular physics textbook entitled *Elementare Lehrbuch der Physik nach den neuesten Anschauungen* (Elementary physics according to the new viewpoints [1895]), which ran to four editions. In 1870, Dressel, accompanied by a group of science teachers, including Theodor Wolf (1891–1924), a Jesuit geologist, traveled to the Republic of Ecuador to take charge of the Polytechnic School founded by Ecuadorian president Gabriel García Moreno (1821–75, in office 1861–65, 1869–75). They returned to Germany six years later after the death of the president and the closing of the school.

In 1884, members of the Jesuit Belgian province, aware of the need to maintain scientific studies as part of Jesuit training, presented a document to the then vicar general Anton Anderledy (1819–92, vicar general 1883–87, superior general 1887–92) for distribution to the other provinces.[13] The document requested the inclusion of physics, chemistry, geology, botany, and zoology in the programs of philosophical studies. The response to this request varied greatly among the different provinces; however, though most of the provinces agreed that sciences had to be taught, the general consensus was that it was impossible to establish general rules for the teaching of science. Similar recommendations, for example those presented by Wulf in 1921, were sent to Rome insisting on the need for Jesuit students to receive proper scientific education. Yet these proposals were not promulgated in any official document, and the teaching of science ultimately remained the responsibility of each Jesuit province. When referring to the training of Jesuits, the Twenty-Seventh General Congregation (1923) repeated the obligation to follow St. Thomas in theology and philosophy (decree 99) and stressed the need for care to be taken with regard to dangerous doctrines (decree 102).[14] Nothing was said, however, about the study of science.

---

11   Theodor Wulf, *Die Faden-Elektrometer* (Berlin: Dümmlers, 1933).
12   Per Carlson, "A Century of Cosmic Rays," *Physics Today* 65 (2012): 30–36.
13   "Examen postulati provinciae Belgicae de studio scientiarum physicarum a patribus eiusdem provinciae institutum (1884)," Archivum Romanum S.I. (Rome), *De scientiis, studia* 4–21 (1884–1921).
14   "Statuta Congregationis Generalis XXVII," *Acta Romana S.I.* 4 (1924): 59–61.

With the reorganization of ecclesiastical studies proposed in 1931 (in the apostolic constitution *Deus scientiarum Dominus* by Pius XI [1857–1939, r.1922–39]), new programs of philosophy and theology were to be established. In 1934, the Jesuits adapted their programs to fit these new norms.[15] Topics related to science—such as the structure of matter, the origin of the universe, and biological evolution—were included in a new, additional subject called "scientific questions related with philosophy." These topics were to be considered in relation to philosophical problems, so that there was no teaching of science in the strict sense. In this subject, "scientific questions" were considered from a philosophical perspective.[16] In this way, the teaching of mathematics and science practically disappeared from the Jesuit faculties of philosophy in the following years. What remained was philosophical "cosmology," taught from the perspective of neo-Thomist doctrine, with the problems discussed in the "scientific questions" also being examined from the point of view of philosophy. This organization of philosophical studies was approved by the Twenty-Eighth General Congregation in 1938.[17]

However, despite this reorganization, after the congregation of 1938 a number of cosmology professors nevertheless began to introduce scientific subjects into their lectures. Peter Hoenen (1880–1961), for instance, who held a doctorate in science from the University of Leiden, initially worked as a physics and chemistry professor in several Jesuit colleges in the Netherlands and introduced these subjects to his cosmology lectures in Valkenburg's faculty of philosophy; his textbook *Cosmologia* (1931) ran to five editions. Similarly, in addition to publishing a textbook on Thomistic natural philosophy, *Cosmologia* (1959), Filippo Selvaggi (1913–95), who taught classes in cosmology and the philosophy of science, also wrote several books on physics, such as *Orientamenti della fisica* (Overviews of physics [1961]) and *Problemi della fisica moderna* (Problems of modern physics [1953]).

These examples notwithstanding, by the second half of the twentieth century it was becoming increasingly clear that cosmology, though a philosophical subject, had to be more open to the problems raised by modern science. Consequently, in 1960, Superior General Jean-Baptiste Janssen (1889–1964, in office 1946–64) issued a letter calling for Jesuit professors and students to receive sufficient scientific training. Thus Jesuit students needed sufficient knowledge

---

15   "Approbantur statuta facultatum theologiae et philosophiae in collegiis Societatis Iesu erectarum," *Acta Romana S.I.* 7 (1934): 606–33.

16   F. M. Palmés, J. Torre Almenara, and J. M. Caballería, "Las 'cuestiones científicas' en las facultades de filosofía," *Estudios eclesiásticos* 55 (1935): 1–32.

17   "Decreta Congregationis Generalis 28, 4: De studiis," *Acta Romana S.I.* 9 (1938–40): 37–40.

of mathematics, physics, chemistry, biology, geography, and astronomy before beginning their philosophical studies. Yet while he was aware that the scientific training acquired at the secondary school level by those joining the Society was insufficient, he gave no solution to this problem.[18]

Due to the lack of specific training in science during philosophical studies, the need for more advanced studies in science meant that, after finishing their philosophical studies, some Jesuit students with a special interest in science were sent to study at Jesuit or secular universities. This special training was reserved for those appointed to become professors of science in Jesuit schools or to work at observatories. The time for this was the period between philosophy and theology, when young Jesuits were usually sent to teach at schools. For those selected, this was a period of so-called "special studies." Another possibility was to undertake scientific studies after finishing theology. The custom of this type of specialized study in science began soon after the restoration of the Society of Jesus. In a letter (1947) addressed to Jesuits about their work, Janssen placed a clear emphasis on academic studies, including the study of the natural sciences: "This applies to sacred disciplines but also to *profane sciences*, which should not be considered as foreign to our vocation in the church or the Society." Among the examples he mentions are archeology, paleontology, and physics. Janssen insisted on the need for serious scientific work in these fields as it would bring prestige to the church: "Today, people do not esteem the theologian or the Catholic philosopher, but would have a great appreciation for the historian, mathematician, and astronomer."[19] Yet despite these recommendations, formal science teaching in Jesuit training was no longer present in the philosophy programs, as we have seen.

In present times, a decrease in the number of vocations and the trend of entering the novitiate at a later age, often after having completed university studies, have radically affected the situation. As a consequence, formal teaching of science is no longer part of the Jesuits' training. Moreover, those who enter the Society having already studied science or engineering are not always encouraged to continue with active work in science. This has contributed to a certain decline in the presence of Jesuits in scientific fields, as we will see in more detail later.

---

18   Jean Baptiste Janssen, "Instructio de cosmologia in nostris scholasticatibus docenda," *Acta Romana S.I.* 13 (1960): 809–14.

19   Jean Baptiste Janssen, "Epistola de ministeriis nostris (22 June 1947)," *Acta Romana S.I.* 11 (1946–50): 299–336.

4   Science in Jesuit Universities, Colleges, and Secondary Schools

As mentioned earlier, the suppression of the Society of Jesus brought with it
the loss of the universities, colleges, and schools the order had managed in
Europe and America since the sixteenth century. With the nineteenth century,
a process of reform began whereby theology either lost its predominant position or in many cases totally disappeared from the curriculum, with much
more emphasis being placed on the natural sciences. The German universities,
beginning with those of Berlin and Göttingen, were pioneers in this process,
which involved giving a major role to research (the so-called Humboldtian
model of higher education). The change to this double orientation of teaching and research was adopted by all modern universities. In France, a radical
reform of all the universities took place during the Napoleonic period whereby
they became directly dependent on the state as a centralized, secular, national
body. A similar process of university dependence on the state then spread to
other countries in Europe, notably Italy and Spain. As a result, it became practically impossible for the Jesuits to take back the universities they once had in
Europe, except for those devoted exclusively to ecclesiastical studies, though
they were eventually able to establish new institutions of higher education in
many European countries as the situation changed over time. In other countries, such as the United States, the situation was different, enabling the Jesuits
to establish institutions of higher education (colleges and universities) from
the nineteenth century onward. The situation was also different in India, as
the Jesuits were able to use the British system to establish colleges associated
with state universities.

Today, there are 190 Jesuit universities spread throughout the world, all of
which were established from the middle of the nineteenth century.[20] In the
United States, where there is a strong tradition of private university education,
the first colleges and universities were established shortly after the restoration.
The first was Georgetown University in Washington, DC, established by John
Carroll (1735–1815), a former Jesuit and the first Catholic archbishop of Baltimore, in 1789, the management of which was entrusted to the Jesuits after the
restoration.

Nineteen of the twenty-eight Jesuit universities in the United States were
founded before 1900. The most well-known are Georgetown (1789), Saint
Louis (1818), Fordham in New York (1841), Boston College (1863), Loyola in
Chicago (1870), San Francisco (1875), Regis in Denver (1877), and Marquette in

---

20   "List of Jesuit Institutions," Wikipedia; https://en.wikipedia.org/wiki/List_of_Jesuit_edu-
        cational_institutions (accessed September 3, 2018).

Milwaukee (1881). Between 1950 and 1960, all Jesuit universities in the United States experienced a rapid growth in students, faculty, and resources.[21]

In India, there are currently sixteen Jesuit colleges, including St. Xavier's in Kolkata (1860), St. Xavier's in Mumbai, (1869), and Loyola in Chennai (1925); in Central and South America, there are twenty-four, including the Pontificia Universidad Javeriana in Bogotá (Colombia) (1623 and 1937), the Pontificia Universidade Catolica de Rio de Janeiro in Brazil (1940), and the Universidad Iberoamericana in Mexico (1943); in Asia, there are eleven Jesuit colleges, including the Ateneo de Manila University (Philippines) (1859), Sofia in Tokyo, Japan (1913), and Sogang in Seoul, South Korea (1960); in Africa and the Middle East, there are five Jesuit universities.

Although there are also nine Jesuit institutions in Europe, few have science programs. In Belgium, one finds in Namur the Facultés Universitaires Notre Dame de la Paix (today the University of Namur), founded in 1831. In France, the Institut Catholique d'Arts et Métiers (ICAM) in Lille was founded in 1898 and has offered higher studies in engineering since 1911. This latter institution in turn served as the inspiration for the Instituto Católico de Artes e Industrias (ICAI), an engineering school the Jesuits established in Madrid in 1908. Its founder, José A. Pérez del Pulgar (1875–1939), who had studied science in Göttingen, introduced the subject of electrical engineering to Spain. In 1978, the ICAI became part of the Jesuit Universidad Pontificia Comillas in Madrid. In the Universidad de Deusto (Bilbao), which was founded in 1886, there are currently some engineering departments, as is also the case in the Universidad Loyola Andalucía (Seville and Córdoba), founded in 2013. The Jesuits also established two laboratories in Roquetas (Tarragona) for studies in chemistry and biology respectively in 1905 and 1908. In 1916, they moved to Barcelona and merged together under the name Instituto Químico de Sarriá (IQS), which amalgamated with other institutions to form the Universidad Ramón Llull in 1990.

4.1   Jesuits Teaching Science in Jesuit Universities

Hence not all Jesuit universities have science programs, and, in those that do, the teaching of science varies greatly, as does the quality of the science departments themselves. In general, Jesuit science professors have tended to focus

---

21   Raymond A. Schroth, *The American Jesuits: A History* (New York: New York University Press, 2007), 174–96; Gerald L. McKevitt, "Jesuits Schools in the USA, 1814–c. 1970," in *The Cambridge Companion to the Jesuits*, ed. Thomas Worcester (Cambridge: Cambridge University Press, 2008), 278–93.

on teaching, though many of them have also carried out some research in their fields of expertise.

The total number of Jesuits teaching and carrying out research in science departments has varied over time and among the different schools. While there was a relatively large number of Jesuits who taught science in the years following the restoration, their presence in such work has diminished more recently, especially from the early 1980s onward, so that at present most science teachers in Jesuit universities are laymen.

To gain an idea of how many Jesuits were involved in scientific work, some numbers for the universities of the United States and India may help.[22] When their numbers were at their highest around 1970, there were some 135 Jesuit professors of science in the United States (where most of the twenty-eight Jesuit universities had science programs at different levels), around three to six in each institution: the highest number was at Boston College (fourteen), then Fordham (eight), and Saint Louis (seven). Jesuit science professors were divided between mathematics (forty), physics (thirty-nine), chemistry (twenty-two), biology (twenty-seven), geology (four), and engineering (two). In the same year, in the sixteen colleges in India, there were a total of sixty-eight Jesuit science professors: mathematics (fifteen), physics (twenty), chemistry (fifteen), and biology (eighteen).

This number has diminished over time not only in absolute terms but also in relation to the total Jesuit faculty.[23] In Boston College, while there were six science professors out of seventy-nine Jesuit professors in 1940, in 1960 the number had increased to ten out of 117 (four physics teachers, two chemistry teachers, two biology teachers, and two mathematics teachers); however, in 2000, they had decreased to three, with three emeriti. In Georgetown University, recent changes have been more abrupt: in 1940, there were five out of a total of fifty-eight Jesuit professors; in 1960, seven out of ninety; and, in 2000, none out of sixty-eight. Similarly, in India, in St. Xavier's College, Kolkata, out of thirty Jesuit professors there were four teaching science in 1960 but only two in 2000. In St. Xavier's College, Mumbai, there were seven teaching science out of a total of nineteen, but only two out of twenty in 2000.

---

22   Information obtained from the survey of the international coordinator of Jesuits working in the positive sciences, compiled by François Russo and Alberto Dou between 1975 and 1985 (Biblioteca Comunidad ICAI, Madrid).

23   Data from the corresponding catalogs of the provinces of the Society of Jesus for the years 1940, 1960, and 2000 (Rome: Archivum Romanum S.I.).

This pattern of decline also applies in Europe. In the Facultés Universitaires of Namur, for example, there were seven Jesuit science professors (two teaching chemistry, two teaching biology, two teaching mathematics, and one teaching physics) out of forty-three Jesuit professors in 1940. This number remained stable, and in 1960 there were still seven. Yet by 2000, there were only two Jesuit professors in biology and another two in what was called "scientific research." In ICAI Madrid, between 1930 and 1980 the number of Jesuit professors of science and engineering remained stable with about ten in total. But the number fell abruptly to four in 1990 and to only one in 2000.

The situation is similar in South America. In the two Jesuit universities in Brazil, the Pontificia Universidade Católica in Rio de Janeiro and the Unisinos (Universidade do Vale do Rio dos Sinos) in São Leopoldo, for instance, there were four professors of science in each in 1975, but there was only one for both by 2005.

These examples show that the period with the highest number of Jesuit science professors in Jesuit universities was between 1960 and 1980, which was also the time when there was the highest overall number of Jesuits. The examples also show that the number of Jesuit science professors in Jesuit universities was fairly stable until around 1980, in some cases reaching up to ten percent of a Jesuit faculty, but fell rapidly thereafter. Their decrease is higher than that of the Jesuit presence in the overall faculty. While many factors have contributed to this rapid decline, one of the most important has been the fall in the number of those joining the Society, especially in Europe and the United States, and a change in apostolic priorities with a strong diminution of those dedicated to science. The general problem of the declining numbers of Jesuits involved in science in more recent times will be discussed in further detail later in the essay.

4.2   Science Teaching in Secondary Schools
The reform of education in most countries in the nineteenth century established three levels of education: the primary and secondary schools and the universities. Most of the Jesuit schools founded after the restoration were at the secondary level (students of fourteen to eighteen years of age), though primary schools were also founded.

The post-restoration history of the Jesuits' schools, which spread rapidly throughout the entire world, is usually divided into three periods. The first, between 1812 and 1917, was marked by a great deal of political instability in which the Jesuits were expelled from several countries, leading to the closure of many of their schools. The second period, 1920–65, was one of rapid expansion that saw the opening of numerous schools in many different countries. The third, 1970 to the present day, has witnessed the closure of some schools due to the decline in the number of Jesuits and radical changes in the organization of most schools.

There are currently 380 Jesuit secondary schools in the world—120 in India and nineteen in the rest of Asia, sixty-two in Europe, fifty-four in Central and South America, fifty-two in the United States, and eight in Africa and the Middle East.[24] In addition to these traditional secondary schools, the Jesuits have also been involved in other educational ventures, such as the networks of Fe y Alegría schools in Central and South America with more than eight hundred thousand students and thirty-eight thousand teachers in eighteen countries, the SAFA schools in southern Spain with twenty-six schools and twenty thousand students, and the school network of Cristo Rey in the United States with thirty-two schools and eleven thousand students. In all these schools, science teaching is present at different levels and varies enormously with regard to the subjects covered.

Initially, a large proportion of the science teaching at these schools was carried out by the Jesuits themselves, but this has changed in more recent times due to the changes mentioned above, and especially the fall in the number of those choosing to enter the order. Between 1965 and 1998, for instance, the number of Jesuit teachers fell from around ten thousand to 4,500, while in many schools the proportion of lay science teachers rose to above ninety-five percent of the faculty; the situation remains the same at the present time, with most science teaching at Jesuit schools being done by lay teachers.

In the early years, the Jesuits installed observatories in some of these secondary schools, such as those at Stonyhurst in the UK, Kalocsa (Hungary), Manila (Philippines), and Belén (Cuba), which carried out astronomical, geophysical, and meteorological observations to a high standard. They also established modern physics and chemistry laboratories and set up museums of natural sciences that were pioneers at the beginning of the twentieth century. Some professors of biology acquired a certain prominence in botany and entomology, as we will see.

4.3   *Science Textbooks*
Following in the tradition of the scientists of the early Society, the post-restoration Jesuits have written a large number of science textbooks for use

---

24   "List of Jesuit Institutions," Wikipedia; https://en.wikipedia.org/wiki/List_of_Jesuit_ educational_institutions (accessed September 3, 2018).

at university and secondary school levels. At the university level, these include: Angelo Secchi, *Lezioni elementari di fisica terrestre* (Elementary lessons of terrestrial physics [1879]); Joseph Bayma, *Elements of Infinitesimal Calculus* (1889); Johann G. Hagen, *Synopsis der höheren Mathematik* (Synopsis of higher mathematics, 4 vols. [1891–1930]); August Bellynck and Égide Pâque, *Cours de botanique* (Course of botany [1899]); Adolf Müller, *Elementi di astronomía* (Elements of astronomy [1904]); Eduardo Vitoria, *Manual de química moderna* (Handbook of modern chemistry [1910–55]); José A. Pérez del Pulgar, *Electrodinámica industrial* (Industrial electrodynamics, 4 vols. [1915–19]); Theodor Wulf, *Lehrbuch der Physik* (Textbook of physics [1926]); Edmund J. Burke, *Lectures in Biology: Zoology* (1928); James B. Macelwane, *Introduction to Theoretical Seismology* (1936); Everett Larguier, *Fundamental Concepts in Mathematics* (1953); Johannes N. Haas, *Physiologie der Zelle* (Physiology of the cell [1955]); and Alberto Dou, *Fundamentos de la matemática* (Foundations of mathematics [1974]).

At the secondary school level, science textbooks written by Jesuit teachers include: Giovanni Egidi, *Compendio di aritmetica ragionata* (Summary of reasoned arithmetic [1881]); Bonifacio Fernández Valladares, *Física elemental* (Elementary physics [1900, 1921]); Ignacio Puig, *Curso general de química* (General course of chemistry [1927]); Franz Heselhaus, *Leitfaden der Biologie* (Guidelines of biology [1928]); Ferdinando Tosarelli, *Curso de biología* (Course of biology [1948]); Julio Henao, *Química orgánica* (Organic chemistry [1963]); Léon Talpain, *Mathématiques: Classe de première D* (Mathematics: First class D [1966]); and Carmelo Oñate, *Física: Preuniversitario* (Pre-university physics [1969]).

5   The New Observatories

As well as their work as science teachers at universities and secondary schools, the Jesuits have also made major contributions to the science of astronomy, a field in which the order has been involved ever since the first years of the Society—one need only recall the names of Christopher Clavius (1537–1612), Christoph Scheiner (1575–1650), and Giovanni Battista Riccioli (1598–1671) in Europe, and those of Matteo Ricci (1552–1610), Johann Adam Schall von Bell (1592–1666), and Ferdinand Verbiest (1623–88) in China. In addition to the work of these and other Jesuits, the order's interest in astronomy was also manifested in the observatories established in a number of Jesuit colleges from around 1630, many of which would go on to make major contributions to numerous aspects of astronomy between the sixteenth and eighteenth centuries.

Given this long-standing tradition, it is far from surprising that many of the Jesuits of the restored Society turned once again to astronomy after 1814 and established observatories in their colleges, faculties of philosophy, and in mission countries. Indeed, the most remarkable feature of the Jesuit dedication to science in this new period was undoubtedly the worldwide establishment of observatories, with a total of seventy-four founded between 1824 and 1957: twenty-six in Europe, twenty-one in the United States, eighteen in Central and South America, and nine in Asia, Africa, and Australia.[25] They were of different types, with some becoming internationally recognized research centers in astronomy, meteorology, or seismology, while others were simply small meteorological or seismological stations.

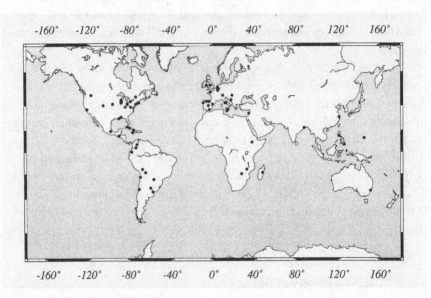

FIGURE 1   Jesuit observatories, 1814–2000

The Jesuit interest in astronomy and geophysics in this new period is also manifested by the names of thirty-one Jesuits who became members of the Royal Astronomical Society of London (RAS) between 1841 and 1977; three were "associate members" (A), and nineteen are recorded in obituaries in the publications of this society.[26]

25   General information on the Jesuit observatories can be found in Pierre de Vregille, "Les observatoires de la Compagnie de Jésus au début du XXᵉ siècle," *Revue des questions scientifiques* 59 (1906): 10–72, 493–579; Agustín Udías, *Searching the Heavens and the Earth: The History of Jesuit Observatories* (Dordrecht: Kluwer, 2003).
26   Agustín Udías, "Serving God and Science," *Astronomy and Geophysics* 42 (2001): 23–24.

TABLE 1   Jesuit membership in the Royal Astronomical Society

Name	Year elected	Country	Observatory
Francesco de Vico	1841 (A)	Italy	Roman College
Alfred Weld	1849	United Kingdom	Stonyhurst
Angelo Secchi	1853 (A)	Italy	Roman College
Stephen J. Perry	1869	United Kingdom	Stonyhurst
Walter Sidgreaves	1891	United Kingdom	Stonyhurst
Aloysius L. Cortie	1891	United Kingdom	Stonyhurst
Edmund A. Goetz	1903	France	Bulawayo
Gustavo Heredia	1904	Mexico	Puebla
Johann G. Hagen	1908 (A)	Germany	Vatican
William F. Rigge	1909	United States	Creighton
Edward D. O'Connor	1914	United Kingdom	Stonyhurst
James P. Rowland	1921	United Kingdom	Stonyhurst
Bernard Swindells	1924	United Kingdom	Stonyhurst
Francis Tondorf	1927	United States	Georgetown
Paul McNally	1929	United States	Georgetown
Daniel O'Connell	1935	Australia	Vatican
William J. O'Leary	1938	Ireland	Riverview
John J. Lynch	1938	United States	Fordham
Patrick Treanor	1947	United Kingdom	Vatican
Francis J. Heyden	1950	United States	Georgetown
Thomas Burke-Gaffney	1952	Australia	Riverview
Joseph Junkes	1952	Germany	Vatican
Richard E. Ingram	1956	Ireland	Rathfarnham
Martin F. McCarthy	1957	United States	Vatican
Pierre L. Gouin	1958	Canada	Addis Ababa
Antony Fynn	1959	Australia	Riverview
Lawrence A. Drake	1965	Australia	Riverview-La Paz
Edmund J. Benedetti	1967	Spain	Vatican
Christopher J. Corbally	1974	United Kingdom	Vatican
Agustín Udías	1977	Spain	

In the table, the Jesuit members of the RAS all belong to observatories (except for the current author). They worked at twelve different observatories and came from eleven countries, although, naturally, the greatest number was from the United Kingdom (nine). The same is also the case with the Jesuits elected members (*non-résident*) of the Académie des Sciences of Paris.

TABLE 2   Jesuit membership in the Académie des Sciences

Name	Year elected	Country	Observatory
Angelo Secchi	1857	Italy	Roman College
Elie Colin	1899	France	Antananarivo
Bonaventure Berloty	1924	France	Ksara
V. Charles Poisson	1935	France	Antananarivo
Pierre Lejay	1946	France	Zikawei
Pierre Teilhard de Chardin	1947	France	

With the exception of Secchi, the Jesuits at the Académie des Sciences were all from France, and they were directors of three mission observatories. The only Jesuit who was not associated with an observatory in the above table was Teilhard de Chardin (1881–1955).

However, despite the large number of Jesuit observatories, it should be noted that the impetus for establishing them did not come from an explicit program mandated by the central Jesuit government in Rome but instead resulted from local and individual initiatives. The attitude of the superior generals in Rome varied—from that of Jan Philipp Roothaan (1758–1853, in office 1829–53), who was opposed to the establishment of the observatory at Georgetown University, to that of Luis Martín (1846–1906, in office 1892–1906), who encouraged the work of Jesuits in observatories. There is only one official document, dating from 1934, which recommended establishing observatories and laboratories in the faculties of philosophy to help teach science to Jesuit students.[27]

The first observatories were astronomical, but they also made other observations such as those of meteorology and geomagnetism. As costs increased for more advanced telescopes, observatories focused on other observations, such as meteorological, seismological, and geomagnetic observations, with more affordable instrumentation.

---

27   "Ad scientias naturales aliasque disciplinas afines docendas instituuntur in collegiis musaea, laboratoria (physicae, chimiae, biologiae, psychologiae experimentalis, etc) atque etaim ubi fieri potest specola astronomica [...]" (To teach the natural sciences and other related disciplines, establish museums, laboratories [of physics, chemistry, biology experimental psychology etc.] and also, where it is possible, astronomical observatories [...], in the colleges); "Statuta Facultatum in Collegiis S.I. erectarum, Titulus V," *Acta Romana S.I.* 7 (1932–34): 630–31.

When the first Jesuit observatories were founded in the mid-nineteenth century, there were few private institutions of this kind, even in Europe and North America; there were practically none in the mission countries of Asia, Africa, and South America, and very few countries had established state meteorological or seismological services. Thus Jesuit work in this area filled an important need, especially in mission countries. The Jesuits working in astronomy, geophysics, and meteorology made important contributions in these fields, where the number of researchers was not then very large. This was especially so in the emerging fields of geophysics and meteorology.

Several factors beyond the desire to further the Old Society's scientific tradition influenced the establishment of the new observatories. One of the most important, as mentioned above, was the need to resist the rationalist current, which had been spreading since the mid-nineteenth century. The scientific work of the Jesuits in their observatories was presented as a practical refutation of such ideas and as a clear proof that Christianity, and the Catholic Church in particular, were not opposed to scientific progress. The Jesuit observatories consequently provided tangible, visible evidence of the church's engagement in science, while their scientific standing clearly demonstrated it was possible to be both religious and a scientist and that science does not necessarily lead to agnostic or atheist positions. The observatories in the mission countries, whose observations were highly appreciated by the scientific community, also served to rectify the practical absence of this type of scientific institution in the countries concerned while providing important practical services such as geodetic measurements for cartography, weather and time services, and earthquake-monitoring.

In more recent times, and especially after 1970, changes in this situation and other considerations have contributed to the decline and closure of many Jesuit observatories or in their management being passed to laymen. One of the main problems faced by Jesuit observatories has been the rising prices for ever more advanced telescopes, and the same also applies for the instruments used in geomagnetic, meteorological, and seismological research. Moreover, in the last two fields, the development of national state services in meteorology and seismology in most countries meant that the work of the Jesuit observatories became redundant. The role of observatories in mission countries was also affected by the new nations in Africa and Asia gaining independence, as the observatories had often been linked to colonial administrations, though, in some instances, Jesuit observatories were transferred to state institutions in the new nations and still continue to operate today. At the same time, Jesuit involvement in observatories and astronomical research more broadly has been affected by changes in the priorities of the Society itself after the Thirty-Second

General Congregation (1974–75), which emphasized the need to engage in social work and the pursuit of faith and justice, and the declining number of
Jesuit vocations in Europe and the United States.

Due to a number of internal and external reasons, therefore, Jesuit observatories experienced a crisis between 1970 and 1980, and most of them
were eventually closed. At present, only four or five continue to operate, but
with a very limited Jesuit presence, save for the special case of the Vatican
Observatory.

5.1   Observatories in Europe and the United States

The first of the new observatories was installed in the Roman College. An observatory had been founded there in 1787 when the college was managed by
the secular clergy; its first director was Giuseppe Calandrelli (1749–1827), a diocesan priest. In 1824, the college was returned to the Jesuits, and the French
Jesuit Dominique Dumouchel (1773–1839), who had studied science at the
prestigious École Polytechnique of Paris, took charge as director. At the time,
it was one of ten observatories in Italy. Dumouchel installed a new telescope
as well as other instruments. In 1835, he was one of the first to forecast the
return of Halley's Comet. Francesco de Vico (1805–48), who had collaborated
with Dumouchel in determining the orbit of Halley's Comet, succeeded him
as director in 1840 and gave an additional impetus to the observatory's work
by launching a project to compile a new catalog of stars, observations of the
satellites of Saturn, and of eight comets between 1844 and 1847. Between 1843
and 1847, de Vico and his assistant Benedetto Sestini (1816–90) made a catalog
of the color of 2,450 stars, separating them into white, blue, yellow, orange,
and red stars. This was a pioneering work, together with that of John Herschel
(1792–1871) and Wilhelm Struve (1793–1864).

Unfortunately, in 1848, the Jesuits were expelled from Rome after the revolution that installed the short-lived Roman Republic (1849). De Vico and Sestini
then moved to Washington, where de Vico took charge of the recently installed
observatory at Georgetown University. De Vico died that same year in London,
where he had gone to buy some instruments.

The Jesuits returned to Rome a year after de Vico's death, and Secchi, one of
the most prominent Jesuit scientists of this new period, was appointed director of the Roman College's observatory.[28] He had spent two years with de Vico
at Georgetown University, where he had worked at the observatory. In 1854,

---

28   Aldo Altamore and Sabino Maffeo, eds., *Angelo Secchi: L'avventura scientifica del Collegio
      Romano* (Foligno: Quarter, 2012); see especially the chapter by Ileana Chinnici, "Il profilo
      scientifico e umano di Angelo Secchi," 43–64.

Secchi moved the observatory to a new site on the roof of the church of Saint Ignatius and installed new telescopes, the main one being a Merz equatorial telescope measuring twenty-four centimeters in diameter.

In the same year, Pope Pius IX (1792–1878, r.1846–78) awarded the observatory the title "Pontifical" and entrusted it with the keeping of official time for the Papal States. But work at the observatory did not last for long: in 1870, the troops of King Vittorio Emanuele (1820–78, r.1849–78) entered Rome, the Jesuits were expelled, and the Roman College and its observatory were confiscated by the state. However, thanks to his international prestige, Secchi was allowed to continue as director until his death. Such a compulsory expropriation (in this case, of the first modern Jesuit observatory) by a government motivated by anti-religious and anti-Jesuit prejudices was part of a pattern that would be repeated in other places.

Secchi was a prolific author, publishing twelve books and more than five hundred articles in scientific journals between 1851 and 1878. He was aware that the future of astronomy lay in the study of the physical nature of the heavenly bodies, or what is now called astrophysics, and not just their position. One of Secchi's most enduring contributions in this field was the first classification of the stellar spectra, which he proposed in 1867. At that time, the study of the spectra of the stars was an entirely new field in astronomy. Secchi divided them into four classes, and this division remains the basis for the present standard Harvard Spectral Classification, a fundamental instrument in astrophysical studies. This is undoubtedly Secchi's most important and lasting scientific contribution, as a result of which he is internationally recognized as a pioneer in astrophysics.

Among Secchi's books, one of the most successful was *L'unità delle forze fisiche* (The unity of physical forces [1864]), which was translated into French, German, and Russian; the book made an important contribution to the search for unity in the nature of physical phenomena, which has been a constant quest in the history of physics, and claims that all physical phenomena can be explained by the general laws of mechanics. Together with many other physicists of the time, Secchi asserted that all forces can ultimately be reduced to some form of movement. In *Le soleil* (The Sun [1870]), originally written in French but followed by an Italian edition, and translated into German and Spanish, Secchi begins by outlining the latest scientific discoveries concerning the Sun before proceeding with his own observations and explanations about solar phenomena such as sunspots, protuberances, and the solar corona. Another of Secchi's books, *Le stelle* (The stars [1877]), which was translated into French and German and summarizes his research on the nature of stars, especially his exhaustive spectroscopic studies, also made an important contribution to astrophysics. During his highly active career, Secchi carried out pioneering research work primarily in the fields of astronomy, meteorology, and geomagnetism. In 1856, he was elected member of the Royal Society of London and the following year of the Académie des Sciences of Paris.

The second Jesuit observatory of this period was established in 1838 at Stonyhurst College. This college was originally founded in 1593 under the royal patronage of Philip II of Spain (1527–98, r.1556–98), in St. Omer, then Spanish Netherlands and now France, for the education of young English Catholics before being moved to England in 1794. From 1843 to 1926, faculties of philosophy and theology for the training of Jesuits were established at nearby St. Mary's Hall. The observatory began its work by making meteorological observations, and in 1845 and 1848 the first Jones equatorial telescopes were installed. In 1846, Alfred Weld (1823–90) was appointed director of the observatory; he launched the observatory's real scientific work with a program of astronomic, meteorological, and geomagnetic observations. In 1860, Stephen J. Perry (1833–89) was appointed director, and he, together with Secchi, is another of the great Jesuit scientists of this period.[29] With Perry as director, the observatory entered its most productive period, serving as the place where many Jesuits who would later become directors of other observatories in different parts of the world received their training. The principal focus of Perry's research was on the relationship between solar activity and the Earth's magnetic field. In recognition of his research on terrestrial magnetism, Perry was elected a member of the Royal Society of London in 1874 (Perry and Secchi are the only modern Jesuits to have received this honor).

Perry's prestigious research on solar activity and terrestrial magnetism led to him being asked to lead four scientific expeditions financed by the British government and the Royal Society. Two of these were to observe solar eclipses and the other two were to observe the transit of Venus across the solar disc. In the first, in 1870, Perry was commissioned by the Royal Astronomical Society to lead one of four groups during the expedition to Cádiz (Spain) to observe the solar eclipse. Perry's most important expedition was in 1874, when he traveled to the Kerguelen Islands in the southern Indian Ocean, near the coast of the Antarctic, to observe the transit of Venus across the solar disc, which was followed by another expedition to observe a second transit of Venus in Madagascar in 1882. Perry also led three further expeditions to observe solar eclipses: in

---

29   George D. Bishop, "Stephen Joseph Perry (1833–1889): Priest, Scientist, Educator" (MSc diss., University of Manchester, 1977), and Bishop, "Stephen Perry (1833–1889): Forgotten Jesuit Scientist and Educator," *Journal of the British Astronomical Association* 89 (1979): 473–84.

the first, in 1886, Perry led a team to the Island of Carriacou in the Lesser Antilles near Grenada; in 1887, he was invited to Russia to observe a solar eclipse in the town of Kineshma bordering the Volga River northeast of Moscow; and his last expedition, in 1889, was to observe the solar eclipse of December 22 from the Salvation's Islands, near the coast of French Guiana. After the observations, Perry died of acute dysentery on board the ship while on the way to Georgetown in British Guiana where he was ultimately buried.

After his death, Perry's work was continued by Walter Sidgreaves (1837–1919), his collaborator and successor as director, who introduced research on solar and stellar spectrography to the observatory. These lines of research were later continued by Aloysius Cortie and Edward O'Connor (1874–1954) until 1932. Cortie carried out studies of solar spectrography during his observations of the solar eclipses of 1905 (Vinaroz, Spain), 1911 (Tonga Islands), and 1914 (Sweden). He also studied the relationship between sunspots and variations of the terrestrial magnetic field. In 1947, due to mounting financial problems, the observatory was closed, and though it reopened in order to make magnetic observations in 1957, the International Geophysical Year, the project did not last long, and in 1974 the observatory was finally closed. At present, college students and amateur astronomers still use the observatory's telescope.

In 1879, Jesuits established an observatory at the college of Kalocsa, Hungary.[30] Cardinal Lajos Haynald (1816–91), archbishop of Kalocsa, who had an interest in astronomy, was behind its founding, and the observatory was named Haynald Observatory in his honor. Carl Braun (1831–1907), who had studied astronomy in Paris, was its first director. He installed an eighteen-centimeter diameter Merz equatorial telescope and a Hilger spectroscope for the study of the Sun. His successor, Gyula Fényi (1845–1927), who was director from 1885 to 1913, focused his investigations on the study of the Sun. His role in Kalocsa was the same as that of Perry in Stonyhurst. Fényi, a tireless observer, made around forty thousand solar observations, of which six thousand were published. His studies centered on the nature of sunspots and solar prominences and fulgurations and their relationship with magnetic storms, publishing some 145 scientific articles on this subject in Hungarian, German, French, and Italian between 1886 and 1925, at a time when the nature of these solar phenomena was still a debated question. Fényi proposed that their origin was due to movement of the material in the solar chromosphere with currents of incandescent gasses passing in different directions. The observatory continued

---

30   Julius Fényi, "Die Sternwarte in Kalocsa, das Haynold observatorium," *Nachrichten der öster-ungarische Provinze* 3 (1906): 14–23. Mojzes Imre, *A Kalocsai Haynald Obszervatórium Története* (Budapest: Országos Műszaki Információs, 1986).

operating after the Second World War, until, in 1950, the communist government confiscated it together with the college.

The observatory at Valkenburg was established as part of the science program in the Jesuit faculty of philosophy.[31] German Jesuits had founded this faculty in 1872 after being expelled from Germany by the Bismarck government (in office 1871–90). Inaugurated in 1896, the observatory was used for both teaching and research. It was situated on a thirty-meter tower with a dome measuring five meters in diameter housing a twenty-three-centimeter diameter Saegmüller equatorial telescope, along with some other minor telescopes and astronomical instruments. The main research carried out at the observatory was the study of variable stars in collaboration with Johann Georg Hagen (1847–1930), director of Georgetown Observatory. The observatory's most important work took place under the directorships of Joseph Hisgen (1868–1910), Alfred Bauer (1872–1912), and Michael Esch (1869–1938). Its observations were published in the seven volumes of *Veröffentlichungen der Sternwarte des Ignatiuskollegs Valkenburg* (Publications of the observatory of Saint Ignatius College, Valkenburg). In the final years, until 1938, Esch worked alone, continuing with the observation of variable stars. In 1941, during the invasion of the Netherlands, German troops occupied the college and the observatory. Nothing is known about what happened to the observatory's instruments; after the war, the Jesuit faculty moved to Frankfurt, but astronomy was no longer taught there.

The observatory of Cartuja was founded in 1902 in the Faculty of Philosophy in Granada, Spain, with three sections for astronomy, meteorology, and geodynamics.[32] The observatory occupied an impressive new building with a large dome measuring eight meters in diameter resting on a circular tower at its center where a Mailhat equatorial telescope measuring thirty-two centimeters in diameter was installed. One of its most prominent directors was José Mier y Terán (1878–1942). The work on astronomy centered on the study of the Sun, but this was soon abandoned, and the observatory instead focused its work on seismology, as we will see later. In 1967, a thirty-two-centimeter diameter Cassegrain telescope provided by Georgetown University, which had closed its observatory, was installed at the Sierra Nevada at a height of 2,506 meters, and relations with several European observatories interested in the altitude and good climatic conditions of the new observatory were established. However,

---

31   Udías, *Searching*, 94–96.

32   Antonio Due Rojo, "El cincuentenario del Observatorio de Cartuja," *Urania* 234 (1953): 67–81; and *Observatorio de Cartuja, 1902–2002* (Granada: Caja de Granada, 2002).

hopes for a renewal of Jesuit work in astronomy soon diminished, and in that year the observatory was transferred to the University of Granada.

At the beginning of the twentieth century, interest in the scientific training of Jesuit students led to the foundation of three laboratories for physics, chemistry, and biology in the philosophy faculty in Roquetas (Tarragona, Spain). The chemistry and biology laboratories began operating in 1904, but in 1914 they moved to Barcelona and became schools for lay students. They merged together and continue to operate today under the name Instituto Químico de Sarriá. The physics laboratory was turned into an observatory dedicated to the study of the relationship between solar activity and terrestrial magnetism on the initiative of Ricardo Cirera (1864–1932), who had worked in the Manila Observatory. The new observatory was inaugurated in 1904 as the Observatorio de Física Cósmica del Ebro[33] and was used to observe the solar eclipse in Spain on August 30, 1905. As we will see below, the observatory made an important contribution to the study of the Earth's magnetic field. Luís Rodés (1881–1939), director between 1920 and 1939, had studied astronomy and astrophysics in the United States at the universities of Harvard and Chicago and had worked at the observatories of Yerkes and Mount Hamilton. Rodés began a line of research on solar and stellar spectrography, solar physics, and the relationship of solar activity to variations in the Earth's magnetic field. He published a popular and accessibly written book of astronomy, entitled *El firmamento* (The firmament [1927]), which presents some of the results of his research. Although the observatory was badly plundered at the end of the Spanish Civil War (1936–39), it was reconstructed under the directorship of Antonio Romañá (1900–81), who managed the observatory between 1939 and 1970 and began a new program of astronomical and geophysical observations. He was an important figure in the recovery of Spanish science in the post-war period. Jesuits continued to direct the observatory until 2000, when it passed into lay hands.

Other European Jesuit faculties of philosophy sometimes installed small astronomical observatories, some of them equipped with a dome and a telescope, to provide demonstrations in lectures. In 1838, for example, the Collège Notre-Dame de la Paix in Namur (Belgium) installed this type of observatory, with a similar one being established in 1884 at the University of Louvain. A small observatory was also established in the faculty of philosophy in Oña (Burgos, Spain) with a telescope located in a small dome on the roof of the main building, which was installed in 1906. This observatory was abandoned when the Jesuits were expelled from Spain in 1931. The Stella Matutina College

---

33   Manuel García Doncel and Antoni Roca, *El Observatorio del Ebro: Un siglo de historia (1904–2004)* (Roquetas: Observatorio del Ebro, 2007).

for lay students had a small observatory installed in 1889; it was used by astronomy professors until 1936. Similarly, the diocesan seminary in Oudenbosch (the Netherlands) had a small observatory installed in 1891. The Jesuits left the seminary in 1929, and the observatory was closed.

The most important Jesuit astronomical observatory in the United States was that of Georgetown University.[34] The observatory was inaugurated in 1844 in a building expressly built for this purpose with a large dome, where, in 1849, an equatorial telescope measuring twelve centimeters in diameter produced by Troughton and Simms was installed. The observatory was the fourth to be set up in the United States, only six years after the first, the Hopkins Observatory in Williamstown, Massachusetts. The first director was James Curley (1796–1889). De Vico also managed the observatory for a time, and Secchi collaborated in the observatory when both were expelled from Rome. The Austrian Johann Georg Hagen was director between 1888 and 1906. He continued the research on variable stars that he had earlier begun in a small observatory he had installed in 1881 in the Jesuit school at Prairie du Chien, Wisconsin. In 1899, Hagen began publishing the results of his research in the first of the nine volumes of the *Atlas stellarum variabilium* (Atlas of variable stars), and he continued to do so after becoming director of the Vatican Observatory in 1906. Paul A. McNally (1890–1955) was director between 1927 and 1948; he took part in several expeditions to observe solar eclipses, for example to Siberia in 1936 and to the South Pacific in 1937. The observatory's last director, Francis J. Heyden (1907–91), tried to give new life to the observatory with a graduate program in astronomy and a series of research projects on solar spectrography, radio astronomy, and the study of planetary atmospheres. Yet, despite these promising projects, in 1972 the university decided to close both the department of astronomy and the observatory. Heyden then moved to the Philippines, where he continued his scientific work at the Manila Observatory for further twenty years.

As was the case in Europe, the Jesuits also founded small observatories in some of their colleges and universities in the United States. As has already been mentioned, between 1881 and 1888 Hagen established one in the high school of Prairie du Chien where he began to observe variable stars. In 1885, Joseph F. Rigge (1842–1913) and his brother William F. Rigge (1857–1927) installed an observatory in Creighton College (Omaha, Nebraska) with several telescopes, and they collaborated with Hagen in the observation of variable stars. In 1895, Jerome S. Ricard (1850–1930) installed a small observatory in the

---

34   Francis J. Heyden, *The Beginning and End of a Jesuit Observatory (1841–1972)* (Manila: Publication of Manila Observatory, 1975).

University of Santa Clara, California, with sections devoted to meteorology and astronomy. In 1900, he installed two telescopes, and in 1928 a new building was added with a large dome measuring fifteen meters in diameter. Today, under the name of Richard Observatory, the building houses an archive on the history of the university. In 1899, the Faculty of Philosophy at St. Mary's, Kansas, installed a small observatory, and one was also installed at Marquette University (Milwaukee, Wisconsin) that had two telescopes. James L. McGeary (1869–1945), the Marquette observatory's first director, made observations of sunspots, comets, and planets, but the university's interest in astronomy gradually diminished, and the observatory, having been used solely for teaching purposes, was finally closed in 1983.

5.2   Observatories in Asia, Africa, Australia, and Central and South America

Toward the end of the nineteenth and the beginning of the twentieth century, the Jesuits also established observatories in mission territories. These observatories had different characteristics from those of Europe and the United States. Rather than having an apologetic aim, they were intended to provide prestige to the missionaries and to help in the scientific development of countries that lacked proper scientific institutions. Observatories thus played an important role in missionary work. Quite often and for long periods, Jesuit observatories were the only scientific institutions that existed in many African, Asian, and Central and South American countries. They were supported by the colonial administrations and provided the international scientific community with valuable observations from places where observations had previously been impossible. They also had a practical function, for example in providing the astronomical basis for time services and for the geodetic measurements necessary for drawing accurate maps of many regions. Their meteorology and seismology departments, in countries prone to frequent hurricanes and earthquakes, provided invaluable practical services. The four observatories that played the most important role in providing these types of services were in Manila (Philippines), Zikawei (China), Antananarivo (Madagascar), and Ksara (Lebanon).[35]

Manila Observatory was installed in the school known as the Ateneo de Manila (today Ateneo de Manila University) by Spanish Jesuits when the country was still a Spanish colony. It opened in 1865, making it one of the first observatories to be established in the Far East, before those of Jakarta and

---

35   Pierre Lejay, "Trois observatoires de mission," Études 182 (1925): 594–602.

Tokyo, and initially made only meteorological observations.[36] Federico Faura (1840–97), who took charge of the observatory in 1867, played a leading role in its beginnings and can be considered its founder. He renewed the instrumentation and began astronomical observations with those of the solar eclipse of 1868. In 1871, Faura traveled to Europe in order to visit the observatories of the Roman College and Stonyhurst. In 1879, he began studying tropical hurricanes (discussed further below). Faura was succeeded as director by José Algué (1856–1930) in 1897, who gave a new impetus to the observatory's work, having previously studied astronomy at Georgetown University where he had worked in the observatory with Hagen. In the same year as Algué's appointment, the astronomical section was enlarged with a new building with a rotating dome measuring ten meters in diameter housing a Merz telescope of forty-eight centimeter diameter. The astronomy section took charge of the time service used for a large part of the Far East, keeping time within an accuracy of one tenth of a second. In 1901, during the US occupation of the Philippines, the United States administration recognized the Jesuits' scientific work in Manila by entrusting the direction of the new Philippines Weather Bureau to the observatory. The observatory thereafter became the center of a network of meteorological stations and produced two lengthy publications about the natural conditions of the Philippines: *El archipiélago filipino* (The Philippine archipelago) and *Atlas de Filipinas* (Atlas of the Philippines), both of which were published by the US government.

Miguel Selga (1879–1956), who had studied astronomy at Harvard and worked in the Lick and Lowell observatories, succeeded Algué as director in 1926, becoming its last Spanish director. From 1920, US Jesuits began to arrive at the observatory. Charles Deppermann (1889–1957), who had gained his doctoral degree from Johns Hopkins University and had worked at Lick Observatory, began working at the observatory in 1926, eventually serving as its director from 1945 to 1957. He played an important role in the study of typhoons, as we will see below. In 1928, William Repetti (1884–1966) arrived in Manila and took charge of the seismology section. Work ultimately ceased at the observatory in 1942, when the Japanese army occupied the Philippines, and the observatory was totally destroyed in a fire in 1945 when US troops were already close to Manila. The fire, which was started deliberately, destroyed the buildings and

---

36   Manuel Saderra-Masó, *Historia del Observatorio de Manila* (Manila: E. C. McCullough, 1915); William C. Repetti, *The Manila Observatory* (Washington, DC: US Government Publishing Office, 1948); James J. Hennesey, "The Manila Observatory," *Philippine Studies* 8 (1960): 99–120. Aitor Anduaga, "Spanish Jesuits in the Philippines: Geophysical Research and Synergies between Science, Education and Trade, 1865–1898," *Annals of Science* (2013), doi:10.1080/33790.2013.841289.

instruments together with the library and its entire collection of astronomical, meteorological, and seismologic observations made over the course of eighty years. Though the observatory was reopened after the war and the independence of the Philippines, the orientation of the observatory changed: the meteorology section, which had been its most important part, was closed due to the founding of the new official Philippines Weather Bureau, and the observatory was organized along research lines in seismology, astrophysics, ionospheric, and solar physics. In 1951, new buildings were built at Mirador, Baguio, and in 1962 at Quezon City. James J. Hennessey (1909–87) became director in 1957 and placed priority on these lines of research, with new instruments for solar physics. As was the case with other Jesuit observatories, work in Manila began to decline in the 1970s. Today, it remains active, though it is devoted mainly to environmental studies. In 2015, the observatory celebrated its 150th anniversary, making it the longest continually operating Jesuit observatory.

When French Jesuits returned to China in 1843, the memory of the great astronomers Schall von Bell, Verbiest, Ignaz Kögler (1680–1746), and Augustin von Hallerstein (1703–74), who had been directors of Beijing's Imperial Observatory during the seventeenth and eighteenth centuries, went with them. The effort to recover this tradition began in 1871 with the establishment of the Observatory of Zikawei (Xujiahui) near Shanghai, which mainly focused on meteorology.[37] Marc Dechevrens (1845–1923), who became its director in 1876, can be considered the true founder of the observatory.[38] He gave it a strong scientific standing with two sections of meteorology and geomagnetism. Stanislas Chevalier (1852–1930), who succeeded Dechevrens as director in 1887, established an astronomical observatory on the Zose (Sheshan) hill, about twenty-five kilometers to the southeast of Zikawei, in 1901. The observatory had a large dome measuring ten meters in diameter housing a Gautier double equatorial telescope measuring forty centimeters in diameter. For twenty-eight years, Chevalier carried out an intense program of astronomical observations, especially of the orbits of asteroids and their perturbations, which he published in seventeen large volumes. The work related to cyclones begun by Dechevrens and continued by Louis Froc (1859–1932), director between 1896 and 1931, will be discussed further below.[39] The observatory had a time service that was highly valued in the Far East. Between 1926 and 1933, the observatory

---

37   *L'Observatoire de Zi-ka-wei: Cinquante ans de travail scientifique* (Paris: Imp. d'Art G. Boüan, 1930).

38   Henri Gauthier, "Un maître en physique du globe (P. Marc Dechevrens)," *Études* 178 (1924): 272–91.

39   Henri Gauthier, "Au service de tous et de chacun: Le pére Louis Froc (1859–1932)," *Études* 213 (213): 273–86.

participated in the international project for the revision of longitudes. In one of the experiments, Zikawei formed a large triangle with Algiers and San Diego to measure differences in longitude, one of which was the difference in longitude between Zikawei and Greenwich with an accuracy of within five-thousandth of a second of arc. In 1934, Zikawei Observatory published the results of this project and gave the determination of the difference of longitude with respect to other observatories and possible sources of errors. One of the aims of the project was to verify Alfred Wegener's (1880–1930) theory of continental drift. However, the project's short duration, less than one year, meant that it was unable to confirm or disprove the proposed motion of the continents. In 1930, at the invitation of the Royal Greenwich Observatory, Zikawei participated in a cooperative international project examining lunar occultation.

In 1931, Pierre Lejay (1898–1958), who held a doctorate from the University of Paris, was appointed director.[40] Lejay introduced new lines of research on atmospheric and ionospheric physics and gravimetry, on which he published a popular textbook, *Développements modernes de la gravimétrie* (Modern developments of gravimetry [1947]). The Japanese–Chinese war and the Japanese occupation of Shanghai in 1938 cut short Lejay's efforts to promote the scientific progress of the observatory. Lejay returned to France in 1939 and continued his scientific work there, but never returned to China. The observatory itself continued operating until December 1949 when it was occupied and confiscated by the communist government of Mao Zedong, providing yet another example of a Jesuit observatory that was forcibly closed in tragic circumstances. The last Jesuit director was the Italian Ernesto Gherzi (1886–1976), whose work on tropical cyclones and microseisms will be discussed subsequently. In 1962, a merger of the Zikawei and Soze observatories reopened under the name Shanghai Astronomical Observatory.

In 1866, the Belgian Eugene Lafont (1837–1908), a professor of science at the Jesuit St. Xavier College in Calcutta (Kolkata), installed a meteorological station to promote science teaching. In 1870, he improved the station with the installation of a Secchi meteorograph. In 1874, the Italian astronomer Pietro Tacchini (1838–1905), director of the Palermo Observatory, who had traveled to India to observe the transit of Venus, persuaded Lafont to establish an astronomical observatory at the college.[41] With Tacchini's help, Lafont installed

---

40   Henri Milloux, "Notice sur la vie et les travaux de Pierre Lejay (1898–1958)," *Institut de France, Académie des Sciences* 7 (1961): 1–7; B. Decaux, "Le R.P. Pierre Lejay (1898–1958)," *L'onde électrique* 284 (1959): 1–4. M. Pierre Tardi, Le R.P. Pierre Lejay: Notice nécrologique," *Revue des questions scientifique* 130 (1959): 1–12.

41   Ileana Chinnici, "An 'Italian Observatory' in India: The History of the Calcutta Observatory," *Studies in the History of Medicine and Science* 14 (1996): 1–21.

a dome in the college with a Merz equatorial telescope measuring eighteen centimeters in diameter. The observatory was used mainly for solar observations, and its researchers collaborated with Alphonse de Penaranda (1834–96) in observing solar eclipses in various places of India. While the observatory eventually closed in 1939, it reopened in 2005 and now focuses on the study of solar physics and the provision of teaching assistance to the college's astrophysics department. New equipment was installed for astrophysical research in 2010, and the observatory was renamed the Fr. Eugene Lafont Observatory.

In 1875, French Jesuits who had been in Syria from 1843 founded St. Joseph University in Beirut, Lebanon, with plans for an observatory. In 1906, the observatory was established in the town of Ksara in the Bekaa Valley.[42] Bonaventure Berloty (1856–1934), who held a doctorate in science from the University of Paris, was its first director, having previously worked at the observatories of Stonyhurst and Ebro. The observatory had four sections: astronomy, geomagnetism, meteorology, and seismology. In 1913, a new telescope was installed in a new tower with a dome. These promising beginnings were cut short, however, by the events during the First World War, when the French Jesuits were expelled from Lebanon and the observatory was plundered by Turkish troops. After the war, Syria became a French protectorate and Jesuits were allowed to return. In 1918, Berloty returned to Ksara, and the observatory was reconstructed with help from the French government. New instruments were installed, among them a new equatorial telescope measuring twenty centimeters in diameter. In 1921, the observatory was entrusted with the direction of Syria's new meteorological service, with fourteen stations in the territory. As in Madagascar, the Ksara Observatory also took an active part in the geodesic measurements for the new maps of Syria. The Jesuits were responsible for measuring the two base-lines and the triangulation of the region.

Berloty was succeeded as the observatory's director by Charles Combier (1880–1950) in 1925, and a new group of Jesuits joined the observatory to take charge of the geomagnetism, meteorology, and seismology sections. The observatory continued its previous scientific work, with the exception of astronomy, which practically ceased after Berloty's death. However, the observatory's work was affected by the independence of Lebanon in 1944, which meant that it could no longer rely on financial and scientific support from France. Its last Jesuit director, Jacques Plassard (1911–2005), was appointed in 1950. Under his management, the observatory focused on geophysical observations and published a long series of observations, along with a climate atlas, catalogs of earthquakes, and different types of maps between 1967 and 1979. No new

---

42   Bonaventure Berloty, "L'Observatoire de Ksara (Liban)," *Ciel et Terre* 23 (1912): 103–8.

Jesuits joined the observatory, and, finally, in 1979 it was transferred to the Lebanese government. A few years later, it was practically destroyed during the war, which affected a large part of the Bekaa Valley. The remainder of the main building is currently part of a winery.

The most important Jesuit observatory in Africa was established in Madagascar[43] by the French government. The French government entrusted the observatory project to the Jesuits, with Élie Colin (1852–1923), a professor of science, being the person chosen to fulfill this task. He prepared for the role by working with Perry in Stonyhurst, and in 1889 Colin arrived in Madagascar and began the construction of the observatory near the capital Antananarivo, on the hill of Ambohidempona, situated on a plateau 1,402 meters above sea level. At that time, it was the highest altitude observatory in the world. The observatory had a large dome measuring eight meters in diameter with an Eichen telescope measuring twenty centimeters in diameter and three other smaller ones; it also had a range of meteorological and geomagnetic instruments. Besides the astronomical observations, the observatory provided the basic information for the first geodetic measurements to be made in order to draw up the maps of the island and provide the time service for the region.

In 1895, relations between the governments of Madagascar and France were broken off, and the Jesuits were expelled from the observatory and the country as a whole. In the ensuing armed conflict, which ended with the French occupation of Madagascar, the observatory was practically destroyed. In 1896, Colin returned to Madagascar and began reconstructing the observatory on a more modest scale; it was inaugurated in 1902. Between 1896 and 1906, Colin, together with Désiré Roblet (1828–1914), established the geodetic baselines for the maps of the region. In 1898, in recognition of this work, they received the Herbert-Fournet Award from the Geographical Society of Paris. In 1923, Charles Poisson (1882–1965), who had been an officer in the French navy and took part in the First World War, succeeded Colin as director. Though Poisson renovated some of the seismological and geomagnetic instruments, and continued the cyclone forecasting service and the time service, the observatory's scientific work began to decline—by 1945, it was clear that the observatory's equipment was now outdated, yet there were no funds available for its renewal. Lack of financing and Madagascar's independence from France in 1960 further complicated the situation. The Jesuits finally left in 1967, and the observatory was

---

43   Charles Poisson, *Un cinquantenaire, l'Observatoire de Ambohidempona, Tananarive, Madagascar, 1889–1939* (Paris: Editions Dillen, 1939); Evelyne Combeau-Marc, "L'observatoire de Ambohidempona (1888–1923): Pouvoir jésuite et science colonial," *French Colonial History* 12 (2011): 103–21.

transferred to the University of Madagascar. In 1994, it was renamed the Institute and Observatory of Geophysics in Antananarivo.

The first Jesuit meteorological observations in Africa were the work of Hungarian Jesuits, who established stations at Borona in 1883 and at Zumbo in 1890, near the Zambezi River in today's Mozambique. Kalocsa Observatory published the results of their observations in its bulletins. The stations were closed in 1905. In 1903, Jesuits established another observatory in Bulawayo, Zimbabwe, the third in continental Africa. Edmund Goetz (1865–1933), a native of Alsace who had joined the Jesuits in England, was its founder. He had previously worked as a professor of science and had worked at the Stonyhurst and Georgetown observatories. The observatory had sections for astronomy, geomagnetism, and meteorology. In 1913, Goetz installed a Cooke equatorial telescope measuring sixteen centimeters in diameter, which he used to observe variable stars in collaboration with Hagen; his most important scientific contribution was in climate studies of the region and in magnetic surveys. However, the Jesuits' work at the observatory did not last long, as in 1926 the observatory was transferred to the colonial administration. At present, it continues to function as a state institution under the name of Goetz Observatory, in honor of its founder, and is mainly dedicated to meteorology and seismology.

The Jesuits also managed another observatory in Africa, which was part of the University College of Addis Ababa, a state institution in the Ethiopian capital.[44] The government of Haile Selassie (1892–1975, in office 1930–36, 1941–74) decided to establish an observatory following the recommendation made during the International Geophysical Year in 1955 for an observatory situated near the magnetic equator. Though it primarily focused on geomagnetism, the observatory also had sections for meteorology and seismology. In 1958, Pierre Gouin (1917–2005), a Canadian Jesuit who was a professor of physics in the University College of Addis Ababa, was appointed its director. Gouin carried out continuous magnetic observations as well as other observations, such as those of the solar eclipses of 1959 and 1961, and field gravity surveys of different regions of Ethiopia. In 1978, the new government of Ethiopia, which had overthrown the emperor, removed the Jesuits from the observatory and expelled Gouin from the country. The observatory continues to operate today as the Institute of Geophysics, Space Science, and Astronomy of the Addis Ababa University. With these observatories, the Jesuits played a vital part in planting the seeds of some of the first scientific institutions in Africa, many of which remain active today, albeit without Jesuit direction.

---

44   Paul Mohr, ed., *Geophysical Observatory of Addis Ababa University: 50th Anniversary Symposium Proceedings* (Addis Ababa: Addis Ababa University, 2007).

In 1907, Edward F. Pigot (1858–1929), an Irish Jesuit first appointed to the Zikawei observatory, founded a small meteorological and seismological observatory at St. Ignatius College in Riverview, New South Wales, Australia, where in 1922 he installed a telescope and began a program for observations of variable stars. The observatory's astronomical work was later continued by William J. O'Leary (1869–1939) and Daniel O'Connell (1896–1982), though it subsequently placed a greater emphasis on seismological research. In 1972, the meteorological station was closed, but the seismological station continues to function today, supported by the Australian government agency Geoscience Australia.[45]

There were only a few Jesuit astronomical observatories in Central and South America, and in general they were relatively small when compared with those in other countries. They concentrated mainly on meteorology and seismology, which is not surprising given the frequent occurrence of tropical hurricanes and earthquakes in the region. Hence, while astronomy was not totally neglected in these observatories, it had a secondary role, except for the observatory at San Miguel in Argentina. Again, little social interest in this type of scientific work and the high cost of telescopes may also have had an influence on the lack of astronomical observatories in these countries.

One of the few specifically astronomical observatories managed by Jesuits in South America, although for only a short time, resulted from the effort of Gabriel García Moreno, president of Ecuador, to modernize the scientific institutions of his country. In 1869, he founded the Escuela Politécnica (polytechnic school) in Quito, taking as a model the École Polytechnique of Paris. To run the school, he asked for the help of the Jesuits, who sent six young German professors of science from Valkenburg. Two years later, García Moreno decided to establish an astronomical observatory and asked Johann B. Menten (1838–1900), one of the German Jesuits, to take charge of the project. Buildings for the observatory were finished by 1875, but President Moreno was murdered in that year and the new government withdrew its support as it identified the Jesuits with the previous president, whose policies it opposed. The Escuela Politécnica was closed, and the Jesuit professors returned to Germany. Menten, however, stayed in Quito, where he left the Jesuit order and remained in charge of the astronomical observatory until 1882. The Escuela Politécnica was not reopened until 1935.

The meteorological observatory of the College of Belén in Havana, Cuba, played a highly important role in the study and forecasting of Caribbean

45   David Branagan, "Earth, Sky and Prayer in Harmony: Aspects of the Interesting Life of Father Edward Pigot, S.J., B.A., M.B., B.C.H. (1858–1929); A Jesuit Seismologist; Part I," *Earth Science History* 29 (2010): 69–99.

tropical hurricanes, as will be discussed in further detail later in the paper. In 1882, Benito Viñes (1837–93), who became its director in 1870, installed a Cooke equatorial telescope measuring twelve centimeters in diameter, which he used for his collaboration with Perry in the observation of the transit of Venus on May 19, 1884. In 1897, the observatory moved to the new building of the college where two towers were added, one with a dome for the telescope.[46] The college was again moved to a new site in 1925, though the telescope was not installed and the observatory's work was restricted to meteorological research.

In Mexico, the Jesuits established two observatories in Puebla and Saltillo. The most important, founded by the Italian Pedro Spina (1839–1925), opened in 1877 in the college of Puebla with sections for astronomy, meteorology, geomagnetism, and seismology. In 1880, the observatory installed a dome with a Dollond telescope, which was replaced two years later with a more powerful one. This was the second astronomical observatory to be installed in Mexico, only two years after the National Observatory of Chapultepec. Gustavo Heredia (1869–1926), director from 1905 to 1910, carried out meteorological, geomagnetic, and astronomical observations. His studies focused on the observation of sunspots and features of the solar chromosphere. The observatory was closed down together with the college in 1914 during the Mexican revolution.

The Jesuits also made meteorological observations in a small observatory in the Colegio San Calixto, in La Paz (Bolivia). The observatory's first director, Pierre M. Descotes (1877–1964), had spent some time at the Cartuja Observatory (Granada) preparing for his work. The observatory's main contribution was in seismology, as we will see later. Descotes installed a small telescope and took charge of the time service for Bolivia, and in 1926 he participated in the international project for the revision of longitudes, proposed by the International Commission of Longitudes. He contributed to this project by accurately determining the geographical coordinates of the observatory through astronomical observations, as a result of which he discovered that the longitude of the observatory of Santiago de Chile had an error of 1.18 seconds. The existence of this error was confirmed by new observations made in the observatories of Santiago de Chile and Buenos Aires. As a consequence of this earlier error, the accepted geographical longitudes of the South American continent had been misplaced by four hundred meters to the west and had to be corrected.

The origin of the Jesuit Observatory of Cosmic Physics in San Miguel, near Buenos Aires, Argentina, lay in the individual astronomical interest of

---

46   Manuel Gutiérrez Lanza, *Apuntes históricos acerca del Observatorio del Colegio de Belén, Habana* (Havana: Avisador Comercial, 1904).

Fortunato J. Devoto (1872–1941), auxiliary bishop of Buenos Aires.[47] The bishop sought to complement the work of the two astronomical observatories that already existed in Argentina with a new observatory devoted to astrophysics, which was to be managed by the Jesuits, who proposed a site near the faculty of philosophy and theology in San Miguel. Ignacio Puig (1887–1961), who was assistant director of the Ebro Observatory, was appointed director of the new observatory, which was inaugurated in 1935. It occupied a new three-story building and was devoted to astrophysics, cosmic rays, solar physics, and terrestrial electromagnetism. Up-to-date instruments were acquired and observations soon began. Puig was a prolific writer of textbooks that helped to popularize a range of different scientific subjects.

After Puig returned to Spain in 1943, Juan Antonio Bussolini (1905–66), an Argentinian, was appointed director. In 1960, Bussolini installed new instruments, such as a solar telescope with Lyot filter, a spectroheliograph, and two radio-telescopes measuring five meters in diameter. A new building was added, and solar physics became the main focus of its research. Bussolini's early death seriously affected the observatory. In 1969, the new director, Mariano Castex, who later left the Jesuits, began a rapid expansion of the observatory, which in the long run proved fatal. New buildings and instruments were added, and the observatory lost its Jesuit character and was renamed the National Observatory of Cosmic Physics. In 1977, the observatory was officially transferred to the Comisión Nacional de Actividades Espaciales (National Commission of Space Activities).

5.3   The Vatican Observatory

The Vatican Observatory has a special place among the observatories managed by Jesuits.[48] It is not strictly a Jesuit observatory, but an observatory entrusted to the Jesuits by the popes. In 1774, during the Society's suppression, an observatory was established in the Roman College that Clement XIV considered a papal observatory, and in 1854 Pius IX gave it the official title of the Pontifical Observatory. Its expropriation by the Italian state in 1870 left the Holy See without its own observatory. At a time when the church was accused of being opposed to science, the popes considered it important to demonstrate that the contrary was true with the work of their own astronomical observatory.

---

47   Ignacio Puig, *El Observatorio de San Miguel, República Argentina* (San Miguel, Argentina: Observatorio de San Miguel, 1935).

48   Sabino Maffeo, *In the Service of Nine Popes: 100 Years of the Vatican Observatory* (Vatican City: Vatican Observatory and Pontifical Academy of Sciences, 1991), and *La Specola Vaticana: Nove papi, una missione* (Vatican City: Specola Vaticana, 2001).

For this reason, in 1890 Pope Leo XIII decided to found a new Pontifical Observatory named the Specola Vaticana (Vatican observatory). The observatory, together with a dome, was built at the Vatican on top of the so-called Tower of the Winds. The Barnabite Francesco Denza (1834–94) was appointed the first director. From its founding, the observatory collaborated with international astronomical projects such as the *Carte du ciel* (Map of the sky) to make a photographic atlas of the sky and a new catalog of stars (astrographic catalog). For the project, a double telescope measuring thirty-three centimeters in diameter was installed in a new dome built on top of the Tower of St. Giovanni on the Leonine Wall.

The Jesuits began participating in the work of the observatory in 1906 when Pope Pius X (1835–1914, r.1903–14) appointed Hagen, the director of Georgetown Observatory, as its director.[49] Hagen focused the observatory's work on astronomy and closed the meteorological and magnetic sections. In 1909, a forty-centimeter diameter Merz equatorial telescope was installed in a new dome on another tower of the Leonine Wall. In 1928, Hagen finished the part assigned to the Vatican Observatory for the stars catalog with entries for 481,215 stars. By the time Hagen died in 1930, the observatory had produced 107 maps for the *Carte du ciel* project in which it had been involved since 1891. Johan Stein (1871–1951), a professor of mathematics and physics in Amsterdam and Hagen's assistant, succeeded him as director of the observatory, which in 1935 was moved to Castel Gandolfo, at the palace of Villa Barberini, and new telescopes were installed on the palace's two domes. Alois Gatterer (1886–1953) began working on stellar spectrography, and in 1933 he installed an astrophysical laboratory. On the occasion of the inauguration of the new premises and instruments, Pius XI formally and definitively entrusted the direction and management of the observatory to the Jesuits.

Ten Jesuits worked at the observatory between 1935 and 1940. During this period, the observatory primarily focused on completing the catalog of variable stars that had begun under Hagen's directorship. The end result, the *Atlas stellarum variabilium* (Atlas of variable stars), which was finally published in 1941, had nine volumes with 579 maps of 384 variable stars and twenty-four thousand reference stars. In 1955, the part of the *Carte du ciel* assigned to the Vatican Observatory was also completed. At that time, the observatory installed a collection of 1,022 meteorites, a gift from the widow of Adrien-Charles,

---

49   H. Dopp, "Johann Georg Hagen, S.J., 1847–1930," *Revue des questions scientifiques* 99 (1931): 5–37; Maffeo, *Specola*, 69–98. Agustín Udías, "The Correspondence of Johann Georg Hagen, First Jesuit Director of the Vatican Observatory," *Journal of Jesuit Studies* 3, no. 2 (2016): 259–78.

marquis de Mauroy (1848–1927), and launched a research project on the structure of the Milky Way. Stein's successor as director, Daniel O'Connell, who had
previously been director of the Riverview Observatory in Australia, renewed
the instrumentation with a Schmidt telescope measuring sixty-five centimeters in diameter, which was installed in a new dome in the gardens of the Villa
Barberini. Under O'Connell's direction, the observatory focused on the study of
binary stars, stellar spectrography, and polarimetry. The next director, Patrick
Treanor (1920–78), who obtained his doctoral degree from Oxford University
and had worked at the observatories of Lick and Yerkes in the United States,
was appointed in 1970. Treanor, who had begun working at the observatory in
1961, researched stellar spectrography and the interstellar medium. After Treanor's death in 1978, George V. Coyne, a former professor at the University of
Arizona, Tucson, became director. In 1980, Coyne, aware of the deteriorating
conditions for astronomical observations in Castel Gandolfo, established an
agreement with the University of Arizona for the joint use of the Steward Observatory, and later to install an observatory at Mount Graham, Arizona, where
conditions for observations are very good, with a new telescope based on advanced technology. Most of the observational work has been moved to this
new observatory, which was inaugurated in 1993. Coyne also sought to engage
in a dialogue between science and religion by organizing a series of international seminars and publications in collaboration with the Center for Theology
and the Natural Sciences in Berkeley, California. In 2006, the Argentinian José
Gabriel Funes and in 2015 the American Guy Consolmagno became directors.
Research at the observatory covers a wide range of subjects, including planetary sciences, astrobiology and exoplanets, stellar astronomy, extragalactic
astronomy, and cosmology. An important aspect of the observatory's work is
the organization of international seminars, workshops, and summer schools.
At present, twelve Jesuit astronomers from six countries are actively engaged
in the observatory's work. They represent the continuation of the long Jesuit
tradition in astronomy.

6   The Earth's Magnetism

A number of Jesuit scientists of the Old Society, including Leonardo Garzoni
(1543–92), Niccolò Cabeo (1586–1650), Jacques Grandami (1588–1672), and
Athanasius Kircher (1602–80), had already engaged in work on the different
aspects of magnetism. Kircher, in particular, devoted a bulky treatise to this
phenomenon, finally comparing God with a great magnet that attracts everything. The Jesuit scientists of the restored Society also had a keen interest in

these phenomena, but now from a purely scientific point of view with a special emphasis on terrestrial magnetism.[50] Their interest coincides with the beginning of the instrumental observations of the geomagnetic field.

In 1858, Secchi installed a complete set of magnetic instruments (magnetometers, inclinometers, and declinometers) in the observatory of the Roman College. For this purpose, a special room built from non-magnetic material was prepared on the roof of the church of Saint Ignatius. Secchi studied the periodic variations of the geomagnetic field and tried to relate them to changes in solar activity, assuming the Sun to be similar to a great magnet. At that time, Edward Sabine (1788–1883), one of the pioneers in the study of terrestrial magnetism, was encouraging other institutions to establish a worldwide network of magnetic stations. He persuaded Weld to begin these observations with a permanent station in Stonyhurst. Weld agreed, and in 1858 he installed a complete magnetic station and began observing the absolute values of the components of the geomagnetic field and the temporal variations. Thus Stonyhurst and the Roman College were among the first geomagnetic observatories in the world. Sidgreaves began carrying out monthly systematic observations of the components of the geomagnetic field in Stonyhurst, which in 1865 Sabine presented to the Royal Society as an example of accurate observations. These observations continued until 1974, when the observatory was closed: with more than one hundred years of observations, the project constitutes one of the longest ever series of magnetic observations to have been taken at the same site. Between 1868 and 1871, Perry, together with Sidgreaves, carried out magnetic field surveys in France and Belgium, which were praised by Georg Airy (1801–92), astronomer royal, as the best magnetic observations for a limited region. In 1879, Braun began magnetic observations in Kalocsa, which Fényi continued until 1913. Braun and Fényi studied the relation of the variations of the geomagnetic field with solar activity.

From its foundation in 1904, the Ebro Observatory prioritized the study of the relationship between solar activity and the variations of the Earth's magnetic field. In 1920, Rodés, as a result of his study of magnetic storms, proposed that the Sun influences the geomagnetic field through the emission of clouds of electrically charged particles. This idea can be considered a precursor of the modern concept of solar wind. In 1950, the observatory renovated its instrumentation and in 1954 installed the first ionospheric sounder in Spain.

---

50   Pierre de Vregille, "Le jésuites et l'étude du magnétism terrestre," *Études* 104 (1905): 492–511; Agustín Udías, "Jesuits, Role in Geomagnetism," in *Encyclopedia of Geomagnetism and Paleomagnetism*, ed. David Gubbins and Emilio Herrero-Bervera (Dordrecht: Springer, 2007), 460–62.

Romañá and José Oriol Cardús (1914–2012) continued Rodés's research into terrestrial magnetism with a particular focus on magnetic storms and periodic variations. Jesuits in the Ebro Observatory continued this research on terrestrial magnetism with renewed instruments between 1960 and 1990. This research continues today but without Jesuit involvement.

Terrestrial magnetism was also a research subject in observatories in mission lands. In 1887, for instance, the Manila Observatory established a magnetic section overseen by Martín Juan (1850–88), who had been trained by Perry in Stonyhurst. The following year, Juan carried out a field survey of magnetic observations covering several islands of the archipelago. He died during the course of his work, but it was eventually completed by Cirera, who continued the research on some coastal areas of China and Japan. Cirera published his observations and those of Juan in *El magnetismo terrestre en Filipinas* (Terrestrial magnetism in the Philippines [1893]). After the war, magnetic stations with modern instruments were installed in Baguio and Davao. In 1952, the observatory began a program of ionospheric studies using an ionospheric sounder in Baguio and installed two more in Balara in 1962. Magnetic observations are still carried out at the observatory to this day.

In 1877, the observatory of Zikawei acquired magnetic equipment, which in 1908 moved to Lukiapang and in 1933 to Zose. The observations were published in a series of thirty-nine volumes, *Études sur le magnétisme terrestre* (Studies of terrestrial magnetism, 39 vols. [1908–32]). In 1932, Joseph de Moidrey (1858–1936) completed one of the first determinations of the secular variation of the Earth's magnetic field for the Far East region. The observatory continued its magnetic observations until it was closed in 1949. Maurice Burgaud (1884–1977), who began working at the observatory in 1924, was the magnetic station's final director.

Magnetic observations were also made in Central and South American countries. Belén Observatory in Havana, for instance, began magnetic observations in 1877, and these continued until 1925. In 1877, the observatory of Puebla, Mexico, also began making magnetic observations, which lasted until the observatory was closed in 1914.

The magnetic observations made by Jesuits in Africa were among the first ever made in that continent. The first observations were made in the observatory of Antananarivo in 1889, where Mascart magnetometers were installed. Though this work was interrupted by the destruction of the observatory in 1896, observations were able to resume in 1902 and continued under Jesuit supervision until 1967. In 1955, Burgaud, who had been expelled from China, took charge of the magnetic section and renovated the equipment, with a proton-magnetometer among other new instruments. The magnetic station continues to function today as a state institution, having been upgraded in 1983 and 1993. Ever since its foundation in 1903, the Bulawayo Observatory has had a magnetic station, which is located in a separate building. In addition, in 1909 and 1914, Goetz carried out two long field magnetic surveys along two lines of 330 kilometers and four hundred kilometers in Zambia and the Democratic Republic of the Congo. These magnetic surveys are among the first to have been made in Africa. As we have seen, magnetic observations were the primary focus of the Observatory of Addis Ababa. During the twenty years it was under Gouin's direction, continuous and accurate observations were made of the Earth's magnetic field. His research focused on the study of magnetic storms and temporal variations of the magnetic field, and between 1958 and 1962 Gouin completed the magnetic map of Ethiopia. His work was interrupted in 1978 when he was removed from his position following a change in government and expelled from Ethiopia altogether.

Pierre Noel Mayaud (1923–2006) obtained his doctoral degree in Paris in 1955 with a thesis on the magnetic activity at the polar region using the magnetic observations made during the French Polar expeditions of 1954–55. A *maître de recherches* at the CNRS in the Institut de Physique du Globe in Paris, he was an expert on the time variations of the Earth's magnetic field, especially on the magnetic indexes K_n, K_s and K_m that describe these variations. These indexes are the subject of his publication, *Derivation, Meaning and Use of Geomagnetic Indices* (1980). He collaborated for a period with Gouin in geomagnetic research in Addis Ababa. From 1978, he wrote on the history of science, especially on the conflict between the new astronomy and the church with his monumental work *Le conflit entre l'astronomie nouvelle et l'écriture sainte aux XVIe et XVIIe siècles* (The conflict between the new astronomy and the sacred scriptures in the sixteenth and seventeenth centuries, 6 vols. [2005]).

7   Jesuit Meteorological Stations

Most of the Jesuit observatories founded in the mid-nineteenth and early twentieth centuries had sections devoted to meteorology, which was then emerging as a modern science.[51] The Jesuit dedication to this science can be explained by the relatively low cost of the instrumentation and by the need

---

51   Agustín Udías, "Jesuits' Contribution to Meteorology," *Bulletin of the American Meteorological Society* 77 (1996): 2307–15. Jesuit observatories at which there were meteorological sections can be found in Udías, *Searching*, part 2, 183–292.

for well-attended meteorological stations at a time when state meteorological services were not yet fully established. Indeed, in most countries, such services only began operating around 1850, and hence the contribution of the Jesuit stations was much appreciated. Jesuits also saw this as a way to contribute to modern scientific work, with its emphasis on the empirical aspects of instrumental observations. It was also a complement to the science lectures given in Jesuit colleges and faculties of philosophy for the training of Jesuit students. Thus, along with sections on meteorology in their astronomical observatories, Jesuits established many meteorological stations, some lasting only a short time, which occasionally substituted and in other cases provided data for the initiating national services.

The Jesuits established some forty-seven meteorological stations, if one includes those in observatories that kept records for sufficiently long periods of time. Geographically, there were seventeen in Europe, six in the United States, fifteen in Central and South America, and nine in Asia, Africa, and Australia. Among them, the most important were the observatories of Belén (Cuba), Manila (Philippines), Zikawei (China), and Antananarivo (Madagascar), which were dedicated to observing, forecasting, and researching tropical hurricanes.

In 1824, the first meteorological station was installed at the observatory of the Roman College with the most modern instruments of that time. Bernardino Gambara (1814–84), who was in charge of the station from 1824, carried out six daily observations, and this work continued under Secchi's direction until 1878. In 1858, Secchi developed a meteorological instrument he called the "Universal Meteorograph," which simultaneously recorded atmospheric pressure, temperature, humidity, and wind velocity and direction, simplifying their correlation. Between 1862 and 1877, Secchi published his works on meteorology in the sixteen volumes of the *Bollettino meteorologico* of the observatory. At Stonyhurst Observatory, meteorological observations began in 1838 in a pavilion built for this purpose in the school garden where instruments were installed. In 1842, Weld began systematic observations, and in 1850 he became a founder member of the British Meteorological Society. In 1866, under Sidgreaves's direction, the UK Board of Trade supplied the observatory with modern equipment that improved the standard of the observations. The same year, at Sabine's recommendation, Stonyhurst was chosen as one of the seven top-class meteorological stations of the British Meteorological Office. From 1932, James P. Rowland (1875–1948) focused the observatory's work on meteorological observations. Rowland, under the name "Wizard of the North," published his observations and weather forecasts in the local newspapers. Although the observatory closed in 1947, it continued sending meteorological observations to the British Meteorological Office until 1960. Continuous observations from 1848 to 1942 were published in the bulletin of the observatory and still constitute a highly valuable series.

Jesuits installed a number of meteorological stations in Spain, all of which began operating before the Central Meteorological Institute was established in 1887, the first in 1860 at the college in Leon. At that time, this was one of the twenty-eight stations that existed in Spain. In 1870, a station was installed at the school of Orduña (Vizcaya), and in 1880 one was also installed at the philosophy faculty in Tortosa (Tarragona), which in 1904 moved to the Ebro Observatory. The same year, a station was installed at the school of La Guardia (Pontevedra), where between 1888 and 1907 Baltasar Merino (1845–1917) carried out early studies of the climatology and storms on the coast of Galicia. The station closed in 1916, when the school moved to Vigo. Other stations were installed in 1882 at the faculty of philosophy in Oña (Burgos) and in 1900 at the seminary and later Pontifical University of Comillas (Santander). In 1902, the Spanish Jesuits proposed establishing a network of meteorological stations in order to take advantage of their schools in different regions. The central station was installed in the school in Madrid. The network was formed of eight to nine stations, which sent their observations to Madrid where a bulletin was published. However, the project did not last long and ultimately came to an end in 1906, although some of the stations continued to function until 1931 when the Jesuits were expelled from Spain.

In 1868, Jesuits established a meteorological station in Italy, called the Osservatorio Tuscolano, at the college of Mondragone (Frascati). Secchi had proposed this idea because he thought these observations necessary for the study of the climatology of the Lazio region. The observatory maintained its observations until 1944. In 1915, another station was installed in the college of Acireale, Catania (Sicily), which remained active until the college was closed in 1978. The Jesuits also established two meteorological stations in Malta, one on the island of Gozo in 1882 in the diocesan seminary that lasted until 1909, and another between 1877 and 1907 in Malta itself at Saint Ignatius College, founded by English Jesuits. This latter station was one of Perry's initiatives, and its observations were published in the Stonyhurst bulletins. French Jesuits installed a station on the island of Jersey, near the coast of France, where from 1880 onward they had a faculty of philosophy after their expulsion from France. In 1893, Dechevrens, after his return from China, moved to Jersey and installed the station in a separate building with a fifty-meter high metallic tower to study wind components at different heights. Dechevrens studied the motion of air masses and the distribution of temperature and atmospheric pressure in relation to height, research he had started in Zikawei. The tower was dismantled in 1923, after Dechevrens's death, but regular meteorological observations continued even during the war when the island was invaded by German troops. In 1954, the Jesuits left the island, but Charles Rey (1897–1981), who had succeeded Dechevrens in 1934 and managed the station during the war, continued to maintain the station until 1979. The station has belonged to the British Meteorological Service since 1945, and it continues to operate today after 120 years of uninterrupted service under the name of St. Louis Observatory, in memory of the Jesuits who managed it for eighty-six years.

In the United States, the Jesuits only established a small number of meteorological stations in colleges and universities. At Saint Louis University, meteorological observations began in 1860 and from 1925 were part of the activities of the Department of Geophysics, which offered courses in meteorology. From 1944 onward, the university has run a graduate program in meteorology at master's and doctorate levels, the only Jesuit university with such a program. In the observatory of the University of Santa Clara, Ricard's main interest was meteorology, and in 1895 he began to keep records. He thought that there was a correlation between weather and sunspots, so that weather could be forecast from solar observations. Ricard studied the climatology of California, especially its rainfall patterns, and published a local weather bulletin, earning him the popular moniker the "Padre of the Rains." In 1895, Frederick L. Odenbach (1857–1933) installed a meteorological station at Saint Ignatius College (today John Carroll University) in Cleveland (Ohio), where he specialized in the study of atmospheric electrical phenomena. Other meteorological stations were also established in 1910 at Marquette University, Milwaukee (Wisconsin), and Spring-Hill College, Mobile (Alabama). In 1955, Gherzi, having been expelled from Zikawei, installed a meteorological station at Jean de Brébeuf College, Montreal (Canada). He continued his work in meteorology there and in 1958 installed an ionospheric sounder. With his empirical approach to science, Gherzi was a pioneer in the study of atmospheric electricity in Canada.

In several Central and South American countries, Jesuits laid the foundations for what would become national meteorological services. In recognition of the Jesuit expertise in this science, the governments of Ecuador and Colombia entrusted the Jesuits with the organization of their meteorological services in 1870 and 1922 respectively. In 1864, the Jesuits began meteorological observations in the Colegio de San Luis of Quito in Ecuador. Seven years later, this became the central station of Ecuador's national meteorological service. In 1870, Menten became director of the service, remaining in this position until 1875 when the station in the school was closed. In 1920, the president of Colombia entrusted Simón Sarasola (1871–1947) with the organization of the national meteorological service. Sarasola had worked at Belén Observatory in Cuba from 1905, and in 1910 he had established a small observatory in Cienfuegos.

The central station of the service was established at the Jesuit college of San Bartolomé in Bogotá. The project called for twenty secondary stations, but only four were actually installed. Sarasola was the first to carry out studies of climatology in Colombia. In 1939, he left this post and in 1943 returned to Cuba as director of the Belén Observatory. In 1941, Sarasola, along with Jesús Emilio Ramírez (1904–81), whose major contribution to science was in seismology, as will be seen later, founded the Instituto Geofísico de los Andes Colombianos (Geophysical Institute of the Colombian Andes), which had a meteorological section. In 1949, Ramírez became director of the national meteorological service of Colombia and played an important role in its early development. The planetarium of the city of Medellín where he was born is named after him.

In 1854, the Jesuits had established a school, the Real Colegio de Belén in Havana, and three years later they installed a small meteorological observatory, called the Belén Observatory, which would become famous for its studies and forecasting of tropical hurricanes, as we will see in more detail below. The Jesuits established two other observatories in Cuba, one in Cienfuegos in 1910 and another in Santiago de Cuba in 1930; however, both were basically meteorological stations that sent their observations to the Belén Observatory. The station in Cienfuegos also had an astronomical section with several small telescopes.

The observatories of Puebla and Saltillo in Mexico also had meteorological stations, both of which sent their observations to Mexico's meteorological service. They closed in 1914. In 1892, the Jesuits began meteorological observations in La Paz, Bolivia, at the observatory of the Colegio de San Calixto, and these observations still continue today. Small meteorological stations were also established by Jesuits in several other countries of Central and South America. In 1851, one was set up in Guatemala, but it only remained open until 1870. The station established in Santiago de Chile was also operational for a very limited period of time, lasting from 1867 until 1872, whereas the station of the college of Santa Fe (Argentina) operated for a slightly longer period, from 1904 to 1967. In 1914, the Jesuits installed a meteorological station in Sucre, Bolivia, with observations continuing to be made there until 1966. In Asunción, Paraguay, the Jesuits kept meteorological records for only a short time, from 1935 to 1940. Part of the reason for the short lives of many of these observatories is that they were established due to the scientific interests of their founders, as we have seen, and many only remained operational while their founders were still alive.

The main Jesuit meteorological observatories in Asia and Africa were in Manila, Zikawei, and Tananarive, all of which will be discussed in further detail below when the paper turns to the Jesuits' studies of tropical hurricanes. In 1866, Lafont had set up the observatory of Xavier College (Calcutta) as a

meteorological station, and in the following year he forecast the arrival of a cyclone before the official meteorological service had done so, thus bringing the
observatory's work to the attention of the broader public. In 1872, the observatory installed one of the few Secchi meteorographs then in existence. Edward
Francotte (1843–1923), who was in charge of the meteorological observations
from 1869 onward, studied the climatology of the region using the long series
of observations kept from 1868 to 1923. However, the meteorological station
was closed two years after his death. From its founding in 1903, the observatory of Bulawayo had complete meteorological instrumentation. Goetz used
the observations made there to carry out detailed studies of the region's climatology, especially its rainfall pattern, which he published as a monograph, *The
Rainfall of Rhodesia* (1909). In 1909, Ksara Observatory in Lebanon also began
making meteorological observations, and in 1921, after the First World War and
while Lebanon was under the French Protectorate, the observatory took charge
of the meteorological service of Syria with a network of fourteen stations in
Syria and Lebanon. Berloty was its first director and remained in that position
until his death in 1934; he was succeeded by Combier and Jean Rey (1896–1979).
Combier published several studies on the climatology of Syria and Lebanon.
After the independence of Syria and Lebanon, the observatory restricted its
meteorological observations to Lebanon. In 1951, Rey published the first plu-
viometric map of Lebanon and continued in charge of meteorological observations until 1975. In 1907, Pigot began making meteorological observations at
Riverview Observatory (Australia), which were continued on a regular basis
until 1972. From its foundation in 1958, the observatory of Addis Ababa also
had a meteorological station, which was under Gouin's supervision until 1978.

As we saw earlier, in 1883 the Hungarian Jesuits made the first meteorological observations in central Africa in the Barona mission near the Zambezi
River, Mozambique. Laszlo Menyhart (1849–97), who had worked with Fényi
in the Kalocsa Observatory, took charge of the observations in 1893 and installed new instruments and a new station in Zumbo; observations from both
stations were sent to Kalocsa Observatory and published in its bulletins. After
Menyhart's death, interest in meteorology declined, but the stations continued
to operate until 1905, when both were closed down.

Thus, as the above examples demonstrate, beginning in 1824, the Jesuits
managed a total of forty-seven meteorological stations scattered throughout
the world, helping in the early development of the atmospheric sciences. Their
observations filled a gap at a time when most countries had yet to establish meteorological surveys and contributed to the many practical aspects of weather
forecasting. Jesuit stations in Latin America, Asia, and Africa were among the
first to be installed, and they played an important role in the establishment of

national weather services. This is a lesser-known aspect of the Jesuit contribution to modern science.

8   Tropical Hurricanes

One of the areas of meteorological research in which the Jesuits made a particularly important contribution was in the study and forecasting of tropical hurricanes. Jesuits working in observatories in the Caribbean region and the coasts of the eastern Pacific and Indian Oceans were particularly active in this type of work, which comprised practical research together with forecasting the arrival of hurricanes and communicating this information to the public, as well as formulating theories about their nature and motion. In many cases, the Jesuits' early work in this area was pioneering, as it took place at a time when these phenomena were not yet fully understood.

8.1   *Benito Viñes and Caribbean Hurricanes*

Each year, a number of tropical hurricanes or cyclones cross the Caribbean region, causing extensive damage and human casualties in their wake. As well as making an outstanding contribution to meteorological science more broadly, Benito Viñes (1837–93), who was appointed director of Belén Observatory in 1870,[52] played an important role in forecasting and studying the characteristics of Caribbean hurricanes over a period of twenty-four years at a time when not much was known about hurricanes. In his first years as director, Viñes gathered all the information obtained at the observatory from the start of regular observations in 1858 until 1871, which he published in 1872. On the basis of these records, he examined the characteristics of the hurricanes that had affected the island of Cuba and the atmospheric conditions associated with them. With great thoroughness, he sifted the accumulated data on the organization, circulation, and movement of Caribbean hurricanes, a subject that had received little attention, with the exception of the early studies of the American and British meteorologists, such as William Redfield (1789–1857), William Reid (1791–1858), and James Espy (1785–1860). Viñes used an empirical approach in his studies, carefully examining the variations in atmospheric pressure,

---

52   Antonio López de Santa Anna, *Contribución a una biografía completa del P. Benito Viñes Martorell, S.I., célebre meteorólogo de las Antillas* (Santander: Hermanos Bedia, 1957); Luis E. Ramos Guadalupe, *Benito Viñes, S.J.: Estudio biográfico* (Havana: Editorial Academia, 1996), and *Father Benito Viñes: The 19th-Century Life and Contributions of a Cuban Hurricane Observer and Scientist*, trans. Oswaldo García (Boston: American Meteorological Society, 2014).

wind direction, and the aspect of the clouds—their form, motion, and height. He published the results of his studies in two works: *Apuntes relativos a los huracanes de las Antillas en Septiembre y Octubre de 1875 y 1876* (Notes concerning the hurricanes of the Antilles in September and October of 1875 and 1876 [1877]) and *Investigaciones relativas a la circulación y traslación ciclónica de los huracanes de las Antillas* (Research on the cyclonic circulation and translation of the hurricanes of the Antilles [1895]), a work written in 1893 and published posthumously. In 1898, the Weather Bureau of Washington published an English translation of the first work, *Investigation of the Cyclonic Circulation and Translatory Movement of the West Indian Hurricanes*. An English-language article summarizing this work was also published in the United States in 1885, which was republished in 1887 and 1889. In 1888, Everett Hayden (1858–1932), chief of the Meteorological Division of the US Navy and editor of *Pilot Charts*, a publication of the Hydrographic Office of Washington, reproduced parts of Viñes's work in a book about hurricanes.

Viñes proposed two general laws for cyclonic circulation. The first is that winds rotate counterclockwise in the northern hemisphere and clockwise in the southern hemisphere; the second is that cyclonic currents are convergent at a lower height and divergent at higher heights. As a consequence of these laws, there are different forms and motions of types of clouds: low clouds, high cumulus, dense cirrostratus, and fine cirrocumulus. To these general laws, he then added six particular laws concerning the movement and translation of cyclones in the Caribbean region. Viñes in turn demonstrated how these laws could be used to forecast the approximate paths of hurricanes depending on the date and location of their occurrence. While he never claimed that the paths of cyclones follow these laws exactly, he believed that he had established general patterns deduced from observations that were followed by most of them. He was also the first to describe some anomalies produced in the trajectories when, for example, two cyclones approach each other, making clear that they never join. He suggested that changes in the motion of cyclones are due to air currents in the upper layers of the atmosphere, and extended their structure in a vertical dimension to great heights, proposing a three-dimensional structure of cyclones. Viñes also suggested that the energy of cyclones was related to the latent heat of condensation liberated by their cloudy external walls and the intensification of precipitation. His continuous and patient observations and his empirical approach, taking into consideration all meteorological variables, put him well ahead of his time with regard to the nature of hurricanes, as was recognized in the May 1889 edition of the *Pilot Chart*: "These important laws, established by the study and long experience of Fr. Viñes, should be thoroughly understood by every navigator and utilized by shaping his course so as to avoid a hurricane." Similarly, in a lecture to the Naval Academy, Hayden spoke of "Padre Viñes, the eminent Havana meteorologist, whose scientific ability, indefatigable energy, and unrivalled opportunities have linked his name so inseparably with all recent advances in our knowledge of these storms."[53]

Viñes applied his knowledge on the nature of cyclones to the practical task of accurate forecasting, thus preventing many casualties and much material damage. He based his forecasts on careful observations, such as on the state of the sea, changes of atmospheric pressure, and on the aspect and movements of high-altitude clouds, especially cirrus, which diverge from the vortices of cyclones. He made his first forecast in September 1875 when a strong hurricane struck Cuba after four years of relative calm, doing so when its vortex was still 550 kilometers away, and accurately predicted that it would pass through Havana two days later; this was the first accurate forecast of the passing of a cyclone through a city. Viñes published his forecasts from then onward, the accuracy of which was highly appreciated by the local authorities and maritime institutions—many lives were undoubtedly saved and material damage avoided as a result.

As he was aware that he could not observe the atmospheric weather and forecast the path of hurricanes in the whole Caribbean region with only one observatory, in 1876 Viñes began establishing a network of stations that would send their observations by telegraph to Havana. The network was made up of twenty stations and supported by a number of maritime and telegraph companies. Seven of them—in Trinidad, Barbados, Martinique, Antigua, Puerto Rico, Jamaica, and Santiago de Cuba—sent observations on a regular basis, whereas the other stations only sent observations during the cyclone season. Viñes also designed two instruments, which he called *ciclonoscopio* and *ciclononefoscopio*, to detect an approaching cyclone by observing the direction of winds and the distribution and orientation of clouds. He never finished their construction, but Faura and Algué in Manila used Viñes's ideas for similar instruments, as we will see below.

In 1893, after Viñes's death, Lorenzo Gangoiti (1845–1922) was appointed director of the observatory, and he continued the observations, studies, and cyclone forecasts carried out by his predecessor. At the end of the Spanish administration of Cuba in 1898, the United States Weather Bureau installed an office in Havana that initially dismissed the work of the Belén Observatory on the grounds that it was not truly scientific. Soon, however, the US Weather Bureau recognized the value of the observatory's work, closed its office in

---

53   Walter M. Drum, *The Pioneer Forecasters of Hurricanes* (Washington, DC: Stormont and Jackson, 1905), 13–14.

Havana, and asked the observatory to send daily telegrams to Washington detailing weather conditions. The next directors of the observatory, Manuel Gutiérrez Lanza (1865–1943) and Sarasola, continued the study of hurricanes, with Sarasola publishing a summary of the observatory's work over the course of seventy years in *Los huracanes de las Antillas* (The hurricanes of the Antilles [1925]); Gutiérrez Lanza produced a similar work, entitled *Génesis y evolución del huracán del 20 de Octubre de 1926 y catálogo de ciclones en la isla de Cuba 1865–1926* (Genesis and evolution of the hurricane of October 20, 1926, and catalog of cyclones in Cuba from 1865 to 1926 [1928]). The observatory was finally closed in 1961 when the government of Fidel Castro (1926–2016, in office 1959–2008) confiscated the college and closed the observatory in yet another example of a Jesuit observatory being closed by a communist government, like those of Kalocsa and Zikawei. The observatory's last director was Rafael Goberna (1903–1985), who had studied at Saint Louis University.

8.2   Typhoons and Cyclones in the Philippines and South China Sea

The Manila observatory had been founded mainly for meteorological observations and was equipped with the appropriate instruments to carry out this work, including, from 1869, a Secchi meteorograph.[54] In 1884, when the Philippines was still a Spanish colony, a royal decree named the observatory as the official center of meteorology for the Philippines with a network of thirteen meteorological stations, all supported with the necessary financial funding. The stations sent three daily observations—six during the typhoon season—and the observatory was in contact with others in Hong Kong, Macao, Shanghai, and Saigon.[55]

The study of "typhoons" (the name given to Pacific cyclones, also known in the Philippines as *baguios*) by Federico Faura began in 1879.[56] In July of that year, Faura made the first forecast of a typhoon's arrival in Manila, which may also have been the first official warning of a cyclone's arrival in the Far East. In November of the same year, and two days ahead of its arrival, Faura warned of

---

54   John N. Schumacher, "One Hundred Years of Jesuit Scientists: The Manila Observatory," *Philippine Studies* 13 (1965): 258–86.

55   The official character of the observatory is of interest because at that time there was no such official meteorological service in Spain itself; the national service was only founded three years later, in 1887.

56   "Necrological Session in Honor of Rev. Fr. Federico Faura, S.J.: Articles by Luis Rodes, Eduardo Fontseré and Eduardo Alcobé," *Memorias de la Real Academia de Ciencias y Artes de Barcelona* 22 (1930): 69–92. Ricardo García-Herrera et al., "Northwest Pacific Typhoons Documented by Philippine Jesuits, 1566–1900," *Journal of Geophysical Research* 112 (2007); doi:10.1029/2006JD007370.

another typhoon threatening Manila, which enabled precautions to be taken, such as closing the harbor. The accuracy of these forecasts enhanced the observatory's prestige, and its weather reports, especially those forecasting the arrival of typhoons, provided a highly valuable service. Faura published his work on typhoons in the Philippines in 1882, *Señales precursoras de temporal en el Archipiélago Filipino* (Precursor signs of storms in the Philippine Archipelago), which highlighted the phenomena that enable forecasts to be made about their path. Faura, who was in contact with Viñes, soon became aware that typhoons in the Philippines were of the same type of meteorological phenomena as hurricanes in the Caribbean, which meant that the same laws Viñes had proposed for hurricanes could also be applied to typhoons. Faura's approach, like that of Viñes, was basically empirical, based on the observations of different kinds of meteorological phenomena made at the observatory. His careful analysis allowed him to put forward some pioneering notions on the importance of the motion of air masses and on the concept of fronts, including their formation and development. Drawing on ideas developed by Viñes, Faura also created an aneroid barometer that he adapted to detect the arrival of cyclones.

José Algué, who succeeded Faura as director in 1897, continued to study and forecast the arrival of typhoons.[57] In 1898, he published *Baguios o ciclones filipinos: Estudio teórico práctico* (Baguios or Philippine cyclones: A theoretical and practical study), a comprehensive work based on the observations made at the observatory from its founding. The Spanish edition was followed by one in French and an English translation in 1904, *The Cyclones of the Far East*, in which Algué describes the main characteristics of typhoons, the areas where they originate, and the trajectories they follow. He identifies two types of typhoons: those with parabolic trajectories, which move clockwise around a center located in the north Pacific; and those with a straight trajectory, which move westward from the east of the Philippines to the southern coast of China. For many years, this was the most authoritative work on typhoons in the Philippines. Algué improved the instrument designed by Faura, adding an adjustable scale and an indicator of the wind direction to the barometer. This instrument, which Algué called a *barociclonómetro*, was sold commercially from 1905 and was widely used by navigators in the Pacific to warn of the proximity of a storm.

When Spain ceded the Philippines to the United States in 1898, the future of the observatory was clouded by uncertainty given the official status it had been granted by the Spanish government. However, after several visits by Algué to Washington, and despite some hesitation on the part of the US administration,

---

57   Ignacio Puig, "El P. José Algué, S.J.," *Ibérica* 833 (1930): 392–98.

the US government eventually acknowledged the work done at the observatory and handed it the direction of the newly created Philippines Weather Bureau in 1901, appointing Algué as director. The observatory thus retained the official status originally given to it by the Spanish administration, and it later enlarged its network of meteorological stations, establishing fifty-one new stations in the archipelago.

Selga continued Algué's work on typhoons. One of his most important contributions was his extensive catalog of typhoons with data stretching from 1348 to 1934, *Charts of Remarkable Typhoons in the Philippines 1902–1934: Catalog of Typhoons 1348–1934* (1935), which contains important historical research and an atlas with maps of the trajectories of the most significant typhoons.

American Jesuits had begun to replace Spaniards at the observatory from around 1920 onward. One of the most notable was Deppermann, who spent some time at the Geophysical Institute of Bergen (Norway) in 1932, where he came into contact with the Bjerknes school of meteorology (named after the Norwegian meteorologist Vilhelm Bjerknes [1862–1951]).[58] After returning to Manila, Deppermann began applying the ideas of the Bergen school to his research on the formation and evolution of typhoons, using the modern ideas of frontology and the analysis of air mass to tropical storms with reference to the genesis and paths of typhoons for the very first time; it was also the first time that quantitative methods had been used in the study of tropical cyclones. Two of his most influential works are *Outline of Philippine Frontology* (1936) and *Some Characteristics of Philippine Typhoons* (1939), published in Manila by the US Bureau of Printing. In his works, Deppermann describes the physical characteristics of typhoons, such as barometric oscillations and waves and the relation between barometric gradient and wind strength. His works also outline the differences between tropical and extra-tropical storms and the use of the motion of cirrus clouds to prognosticate the movement of cyclones. He also proposed a theoretical model of typhoons in terms of a Rankine vortex with two convection rings.[59] Deppermann based his studies on a broad collection of observations. Considered one of the leading authorities on tropical cyclones, Deppermann and his studies cast a new light on the characteristics of typhoons, their nature, and dynamics, and the value of his work has been recognized by other researchers. In the *Compendium of Meteorology* (1951) edited

---

58   James J, Hennessey, "Charles Deppermann S.J.: Philippine Scientist," *Philippine Studies* 5 (1957): 311–35.

59   A Rankine vortex (named after William J. M. Rankine [1820–72]) is a circular flow in which an inner circular region around the center is in rotation while the outer region is free of vorticity.

by Thomas F. Malone (1917–2013), for instance, Deppermann is cited more than forty times; in his chapter on tropical cyclones, Gordon E. Dunn (1905–94) asserts that tropical meteorology is largely indebted to Deppermann's detailed description and analysis of typhoons in the Philippines. In the chapter on aerology and tropical storms, Herbert Riehl (1915–97) comments that, before Deppermann, meteorologists had largely contented themselves with the application of simple hydrodynamics; other authors in the same book similarly praise Deppermann's contributions to the study of tropical storms.[60]

Despite the valuable meteorological observations made at the Manila Observatory, its work in this area came to an end shortly after the Philippines became an independent state in 1946, as the new Philippine government established a meteorological service that did not rely on the observations made at the Manila Observatory. The observatory consequently closed the section on meteorology, instead choosing to focus on seismology, ionospheric studies, solar physics, astrophysics, and more recently on environmental studies, as we saw earlier.

As has already been mentioned, the observatory at Zikawei was established largely with the aim of carrying out meteorological research and was equipped from its foundation with a complete set of meteorological instruments, including a Secchi meteorograph (a second was added in 1881). Dechevrens gave the observatory a solid scientific standing in meteorology, began studying typhoons, and established an active research network with other observatories throughout the world. In 1879, he published a detailed study on the nature and effects of a typhoon that had caused severe damage in Shanghai, which helped bring the observatory's work to public attention. Thus, the study and forecasting of typhoons, which devastate the coasts of China each year between July and September, became the main focus of the observatory. Although considered a private institution, from 1876 the observatory was associated with the Chinese Maritime Customs Administration. This collaboration, as well as its collaboration with the telegraph and maritime companies, turned the observatory into a true public meteorological service—in fact, it was the only one in operation in China with connections with the other meteorological stations in the Far East. In 1882, Zikawei Observatory began publishing a daily bulletin in the newspapers of Shanghai detailing the weather conditions with a forecast for the coming days, and from 1890 the observatory began to issue synoptic weather maps for the whole coast of China and to send warnings of storms

---

60   Thomas F. Malone, ed. *Compendium of Meteorology* (Boston: American Meteorological Society, 1951). Especially the chapters by Gordon E. Dunn, "Tropical Cyclones," 887–901, and Herbert Riehl, "Aerology of Tropical Storms," 902–16.

by telegraph. Yet despite these achievements, Dechevrens's proposal for a col-
laborative network between all the meteorological stations in China, with the
network centered in Zikawei, was rejected by the British authorities in Hong
Kong. Nevertheless, in 1883 Dechevrens was able to contribute to the practical
aspects of meteorology by developing a code of visual signs to warn mariners
of weather conditions, especially during the typhoon season. A mast was installed for this purpose in the harbor of Shanghai, where signs were displayed,
linked to the observatory by a telegraph line.

Continuing the work of Faura and Algué, Dechevrens began his study of typhoons using the long series of observations available in Zikawei. Drawing on
these records, he developed his own theory about the structure of cyclones that
emphasized the vertical distribution of temperature and of the strength and
direction of winds and their effect in the formation of cyclones. He suggested
that cyclones are formed on the boundary between high- and low-pressure
centers by a vortex of turbulence in the high regions of the atmosphere, between five hundred and seven thousand meters. This attracts to its center the
air masses from below. From this level, air is dispersed horizontally, and at a
certain distance it descends to converge to form the storm center. Dechevrens
called attention to the importance of winds in the high atmosphere and the
formation of clouds at those levels. He related both phenomena with changes
in the path of cyclones, as Viñes had earlier done.[61] Together with Faura and
Algué, Dechevrens was one of the first to study the nature and characteristics
of typhoons in the Far East. A particular focus of Dechevrens's research was
the vertical component of winds, a phenomenon he studied over many years,
and designed an instrument for this purpose called a "clino-anemometer,"
which was installed on top of the Eiffel Tower during the 1889 World Fair in
Paris. Dechevrens had earlier visited France in 1887, traveling from there to the
Island of Jersey, where the French Jesuits had a college, and helped to install an
observatory with a fifty-meter high tower that he used to continue his study of
the vertical component of winds.

Froc, Dechevrens's successor, perfected the system of storm signs to warn
mariners of weather conditions, and from 1896 this new system was put in
place in all of the most important ports of China. In 1915, Froc proposed that
a cannon be fired in the harbor of Shanghai in addition to the visual system
in order to mark the imminent arrival of a typhoon as indicated by the observatory. The system was enlarged in 1918 to include more information and additional illuminated signs that were visible during the night; the same system

---

61   Marc Dechevrens, *Sur la variations de temperature observé dans les cyclones* (Shanghai:
      Observatoire de Zikawei, 1887).

was also adopted in many ports of the Far East coast. The occurrence of more than thirty cyclones each year on the coasts of China indicates the great need for such a system, which helped to prevent many casualties, the loss of ships, and material damage.

The Zikawei Observatory was in continuous contact with other observatories, with shipping companies, and later, by radio, with ships sailing through nearby waters. Froc became so famous for his accurate forecasts and typhoon warnings, sometimes two days ahead of their arrival on the coasts of China, that he was given the name *Père des Typhons*. Froc collected the observations made at the observatories into an atlas with the trajectories of 620 typhoons for the period from 1893 to 1918.[62]

Gherzi, who arrived in Zikawei in 1920, took charge of the meteorology and seismology sections from 1940.[63] In 1927, he began to give meteorological information by radio and kept in contact with ships passing through nearby waters, warning them about storms. Despite the difficulties created by the Japanese occupation and the war between Chinese nationalists and communists, Gherzi was able to keep the observatory in working order and continue his predecessors' studies of typhoons. He was especially interested in their surface characteristics and elaborated some new hypotheses about their formation.[64] He also kept records of the observations made by the ionospheric sounder installed by Lejay, of solar radiation, and of the amount of ozone in the atmosphere. Gherzi's research sought to find a correlation between variations of the ionospheric layers and the motion of air masses that might allow for better forecasts of the weather, and he also tried to find a correlation between storms in the atmosphere and the generation of microseisms in the Earth's crust, as will be discussed later. After the occupation of Zikawei by the communist government, Gherzi moved to Macao, where he reorganized the meteorological service between 1950 and 1953 and published an extensive work in two volumes, *The Meteorology of China* (1951). Gherzi moved again in 1954, first to Saint Louis University and later to Montreal, where he continued his work in meteorology and ionosphere physics.

Tropical cyclones similar to those in the Caribbean and the Pacific are also frequent occurrences in the Indian Ocean, often causing damage in Madagascar and on Africa's east coast. In general, they are produced in the ocean region

---

62   Louis Froc, *Atlas of the Tracks of 620 Typhoons* (Shanghai: Observatoire de Zikawei, 1920).
63   Jean-François Gauvin, "La physique du globe tous azimuts: La carrière scientifique d'Ernesto Gherzi, S.J., 1910–1973," *Archivum historicum Societatis Iesu* 72 (2003): 45–80.
64   Ernesto Gherzi, "On the Constitution of Typhoons," *Bulletin of the American Meteorological Society* 19 (1939): 59–66.

to the east of Madagascar between October and March and propagate from
east to west in a nearly straight path. From its foundation in 1889, the observatory of Tananarive had paid special attention to the study of cyclones and
forecasting their arrival on the island. The observatory installed a network of
secondary meteorological stations throughout the island, later increasing their
number from eleven to twenty-five. These stations depended on the observatory until 1920, when they were passed to the colonial administration. Colin
published his first study on the cyclones of December 9 to 12, 1902 and continued to publish reports on the most important cyclones affecting the coasts of
Madagascar. In 1905, he published a general work on the cyclones of this region,
Les cyclons à Madagascar (The cyclones in Madagascar), in which he observed
that the path of these cyclones generally followed three courses: at a tangent
to the eastern coast, describing a parabola, and penetrating into the interior
of the island by descending through the Mozambique Channel and entering
via the southwest coast of Madagascar. He found that when cyclones enter the
high plateau of the island at altitudes of one thousand and two thousand meters, they rapidly lose strength and the wind velocity diminishes radically.[65]

Poisson continued the observatory's studies of cyclones, observing and forecasting them and taking note of their path. After a cyclone had caused severe
damage in 1927, the French colonial administration appointed Poisson the
technical director of meteorology in Madagascar. In 1931, he published *Les cyclones tropicaux* (The tropical cyclones), which received an award from the Colonial Academy of Science. From 1903 to 1943, the colonial government placed
the observatory of Tananarive in charge of forecasting the arrival of cyclones
in Madagascar. This service and the time service were the only ones supported
economically by the colonial authorities, so the observatory had to focus on
these practical matters. However, the observatory never had sufficient funds
to update and renew its instruments, and the work was largely limited to making observations. The Jesuits consequently found it difficult to continue their
scientific research in Madagascar, except for the study of cyclones, owing to
lack of funds and personnel; Colin and Poisson largely worked alone, and they
were subject to the routine demands of observations with great difficulty in
establishing contact with other scientists.

As well as observing and forecasting tropical cyclones, Viñes, Faura, Algué,
Dechevrens, and Deppermann also proposed some pioneering theories about
the nature of their formation and path of motion, making truly original scientific contributions to the understanding of these phenomena at a time when

---

65   Elie Colin, *Les cyclones à Madagascar* (Tananarive: Imprimerie Officielle, 1905).

they were not yet fully understood. Hence the Jesuits' work in this area cannot be dismissed as one of purely routine observation.

9 Earthquakes and Seismology

Modern Jesuit scientists also had a particular interest in seismology, a science that studies earthquakes, their causes, and the propagation of seismic waves in the interior of the earth; indeed, the Jesuits had such an extensive involvement in this science that it has occasionally been called "the Jesuit science."[66]

Seismology itself is a relatively modern science, dating from toward the end of the nineteenth century with the development of the first seismographs, instruments made to detect and record the ground motion produced by earthquakes, the first of which were developed in Italy and Japan, two seismically active countries at the end of the nineteenth century. Between 1880 and 1920, the first seismological stations equipped with seismographs were established throughout the world. Thus the restoration of the Society coincided with the beginning of seismology as an instrumental science, and its novelty, similar in that sense to meteorology, attracted the interest of Jesuit scientists. Thus, as we will see below, Jesuits began to install seismographic instruments in the new observatories described above and established new seismographic stations in many parts of the world.

9.1 Seismographic Stations in Europe

The first seismographic instrument in a Jesuit observatory was installed in 1888 by Giovanni Egidi (1835–97) in the Osservatorio Tuscolano in Frascati. Egidi constructed this instrument, which he called a "seismoscope." He collaborated with the Italian seismologist Michele Stefano de Rossi (1834–98) in keeping seismological observations that were published in the journal *Bollettino del vulcanismo italiano* (Bulletin of Italian volcanism). In 1908, the observatory installed two new early Cancani and Agamennone seismographs, provided by the Ufficio Centrale di Meteorologia e Geodinamica (Central office of meteorology and geodynamics).

---

66   Manuel M. Sánchez Navarro-Neumann, "Os jesuitas e a sismologia," *Brotéria* 24 (1937): 145–51; Daniel Linehan, "Jesuits in Seismology," *Earthquake Information Bulletin* 16 (1984): 156–65; Agustín Udías and William Stauder, "The Jesuit Contribution to Seismology," *Seismological Research Letters* 67 (1996): 10–19; John Ebel, "The Jesuit Science," *Company* (Summer 2009): 28–31; Agustín Udías, "Jesuits' Studies of Earthquakes and Seismological Stations," in *Geology and Religion: A History of Harmony and Hostility*, ed. Martina Kölbl-Ebert (London: Geological Society S.P., 2009), 135–43.

In 1902, the Cartuja Observatory (Granada) installed two Vicentini and Stiattesi seismographs, some of the first in Spain; as the observatory was located in the most seismic area of Spain, the work of the seismological station soon acquired great importance. Manuel Sánchez Navarro-Neumann (1867–1941), who is regarded as a pioneer in Spanish seismological research, took charge of the station in 1908 and was director of the observatory between 1915 and 1931.[67] As lack of funds prevented the acquisition of new instruments, he designed and built them himself, following the models produced in Germany by Carl Mainka (1874–1943) and Emil Wiechert (1861–1928), building a mechanical seismograph with a mass weighing three thousand kilograms and a ground amplification of one thousand, a great achievement for an instrument of that type. In 1924, Sánchez Navarro-Neumann also built two electromagnetic seismographs of the type newly developed by Boris B. Golitsyn (Galitzin) (1862–1916) in Russia. His principal seismological work was the study of earthquakes in southern Spain, research that he used to produce the first modern catalog of earthquakes in the Iberian Peninsula between 1917 and 1921. Aware of the lack of seismological literature in Spanish, he published the first modern book about earthquakes and their effects on buildings, *Terremotos, sismógrafos y edificios* (Earthquakes, seismographs, and buildings [1916]). His successor as director, Antonio Due Rojo (1898–1975), tried to keep the station in good working order in the difficult years after the Spanish Civil War (1936–39), continuing the work of Navarro-Neumann and publishing studies of the seismicity of southern Spain. However, continued lack of funds prevented the renewal of the instruments, and the development of the State Seismological Service diminished the importance of the observatory's work. Due Rojo was the last Jesuit to take care of the station. In 1904, a seismological station had also been established at the Ebro Observatory with Vicentini seismographs. Although the instrumentation has been continually renewed and the station continues in operation, no Jesuits are currently involved in the station's work.

One of the more unusual Jesuit seismological stations was installed in Ireland in 1908, a country where there are seldom any earthquakes. The station, at Mungret College, near Limerick, had two Bosch-Omori horizontal seismographs. In 1911, William J. O'Leary (1869–1939), a professor of physics, built a seismograph of his own design consisting of an inverted pendulum with a mass of six hundred kilograms held by three wires. In 1915, O'Leary moved his

67   Antonio Due Rojo, *Labor científica del R.P. Manuel Sánchez Navarro, S.J.* (Granada: Observatorio de Cartuja, 1940), and Due Rojo, "El R.P. Manuel M. Sánchez Navarro, S.J.," *Boletín Real Sociedad de Historia Natural* 39 (1941): 117–20.

seismograph to the Jesuit Faculty of Philosophy in Rathfarnham Castle where he built a new instrument of the same type with a larger mass of 1,500 kilograms. O'Leary then moved to Australia in 1929 where he became director of the Riverview Observatory, with Richard E. Ingram (1916–67) becoming director of the seismological station of Rathfarnham Castle in 1935. Ingram had studied mathematics at University College in Dublin, obtained his doctoral degree in mathematics from Johns Hopkins University (Baltimore), and trained in seismology at the California Institute of Technology (Pasadena).[68] He renewed the instruments of the station, but he was more interested in mathematics and the theoretical aspects of seismology, publishing several articles on seismic wave propagation and the mechanism of earthquakes. Ingram was a professor of mathematics at University College Dublin and contributed to the promotion of mathematical teaching in Ireland. He took part in preparing an edition of the complete mathematical works of William R. Hamilton (1805–65). In 1962, he moved to Georgetown University as professor of mathematics and collaborated with the US Coast and Geodetic Survey and the Dominion Observatory of Canada in solving seismological problems. After his death in 1967, the seismological station at Rathfarnham Castle was closed.

A Milne horizontal seismograph had been installed at Stonyhurst Observatory in 1909, and new instruments were acquired in 1920 and 1924: a Wiechert horizontal seismograph and two Milne-Shaw horizontal seismographs. Work in seismology was reduced to recording observations, and the station was closed in 1947. Similarly, in 1920, a Mainka seismograph was installed in the observatory on the Island of Jersey that operated as long as the Jesuits were there (i.e., until 1979). Between 1924 and 1928, the college-seminary of Travnik (Bosnia) run by the Jesuits installed a Conrad horizontal seismograph.

In conclusion, although Jesuits established and managed several seismological stations in Europe, real seismological research was only carried out in the Cartuja Observatory, primarily by Sánchez Navarro-Neumann, who was a true pioneer of seismological research in Spain. Although the Jesuits had up-to-date seismological instrumentation at the Ebro Observatory, their work never went beyond recording observations, and Ingram's theoretical seismological work was not really connected to the seismological station at Rathfarnham Castle.

---

68   Thomas Murphy, "The Rev. R. E. Ingram, S.J.," *Quarterly Journal of the Royal Astronomical Society* 9 (1968): 226–27.

9.2 *The Jesuit Seismological Association and Jesuit Seismologists in North America*

In 1900, Odenbach, who had established a meteorological station in Saint Ignatius College (today John Carroll University) in Cleveland, installed two seismographs of his own design, partly to promote the work of Jesuit colleges and universities in light of the publicity generated by newspaper reports of earthquakes. At that time, Catholic educational institutions in the United States were often viewed with suspicion, and Odenbach consequently believed that their prestige could be enhanced by engaging in this type of work. He conceived the idea of creating a seismological network covering the United States, taking advantage of the geographical distribution of Jesuit colleges and universities, at a time when such a network did not exist. Odenbach presented his idea to the Jesuit superiors and presidents of colleges and universities in 1908, insisting on the great benefit the network would provide to the Society of Jesus and to the Catholic Church. From the apologetic point of view, as with much of the Jesuits' scientific work, the project would have the added benefit of showing that the church was not an enemy of science and progress. Fifteen Jesuit universities and colleges eventually joined the project: Cleveland (Ohio), Georgetown (Washington, DC), Brooklyn (New York), Fordham (New York), Worcester (Massachusetts), Buffalo (New York), Spring Hill (Alabama), New Orleans (Louisiana), Saint Louis (Missouri), Chicago (Illinois), Milwaukee (Wisconsin), St. Mary's (Kansas), Denver (Colorado), Santa Clara (California), and Spokane (Washington). The Canadian college of Saint Boniface (Manitoba) also took part in the project, and thus the stations almost covered the whole continent of North America. Recognizing the need for uniform instrumentation, Odenbach selected the German Wiechert horizontal seismograph, a modern instrument at an affordable price, for use in all of the stations.

Instruments were installed in the sixteen stations in 1911, and the network began working under the name Jesuit Seismological Service (JSS).[69] The network's stations sent their observations to the central station established in Cleveland, which would then process the data and send them to the International Seismological Center (Bureau Central International de Seismologie), which had been established in 1906 in Strasbourg, France. The JSS was the first

---

69   Nicholas H. Heck, "The Jesuit Contribution to Seismology in the U.S.A.," *Thought* 19 (1944): 221–28; James B. Macelwane, "The Forerunner: The Jesuit Seismological Service," in *Jesuit Seismological Association, 1925–1950: Twenty-Fifth Anniversary Commemorative Volume*, ed. James B. Macelwane (St. Louis: Saint Louis University, 1950), 3–16; Udías and Stauder, "Jesuit Contribution," 1516; Carl-Henry Geschwind, "Embracing Science and Research: Early Twentieth-Century Jesuits and Seismology in the United States," *Isis* 89 (1998): 27–49.

seismological network to be established at continental level with uniform instrumentation. At that time, the federal government of the United States did not have a similar seismological network, and there were only forty-four stations in the whole country, which meant that Jesuit stations comprised one-third of the total. However, this auspicious beginning quickly floundered as problems arose in the collaboration between the stations and the government institutions that had promised them funds. Odenbach, who had opposed collaborating with government institutions but was overridden by others—especially Francis Tondorf (1870–1929), who was in charge of the station at Georgetown University[70]—resigned as coordinator at the end of the same year, and proposed that the stations continue working independently and send their observations directly to the center in Strasbourg. This was the end of this first attempt.

A second attempt to maintain a seismological network was made when James B. Macelwane (1883–1956) returned to Saint Louis University in 1925 and became the director of the university's Department of Geophysics, the first such department in a North American university.[71] Some of the Jesuit seismological stations had ceased working, but most, like those of Denver, Georgetown, Spring Hill, and Saint Louis, continued their work, publishing bulletins with their observations. The station at Georgetown under Tondorf's direction had updated its instruments with modern Galitzin electromagnetic seismographs, recently developed in Russia. Sánchez Navarro-Neumann from the Cartuja Observatory encouraged Macelwane to go ahead with the project, which he considered very important. Several institutions in the United States, such as the Carnegie Institution and the National Research Council, also supported the idea, and in 1925 the project received the approval of the Jesuit provincial superiors. The network was established under the new name Jesuit Seismological Association (JSA), with a central station at Saint Louis and Macelwane as president.[72]

The network comprised fourteen stations: Saint Louis, Cleveland, Georgetown, Brooklyn, Fordham, Worcester, Buffalo, Cincinnati, Spring Hill, New Orleans, Chicago, Milwaukee, St. Mary's, Denver, Santa Clara, San Francisco,

---

70   Geschwind, "Embracing Science," 38–42.

71   James B. Macelwane, "The Jesuit Seismological Association," in Macelwane, *Jesuit Seismological Association*, 17–23.

72   Perry Byerly and William Stauder, "James B. Macelwane, S.J., September 28, 1883–February 15, 1956," *Memoirs of the National Academy of Science* 31 (1958): 254–81; Victor Blum, "Sketch of the Life of James Bernard Macelwane, S.J.," *Earthquakes Notes* 27 (1956): 9–11; Henry Birkenhauer, "Father Macelwane and the Jesuit Seismological Association," *Earthquakes Notes* 27 (1956): 12–13.

and Spokane. The central station at Saint Louis not only collected the data from the stations but used these observations and those from other stations to determine the origin times and locations of earthquakes throughout the world. These findings were published in a bulletin that was widely read by seismologists worldwide. The JSA established an agreement with the US Coast and Geodetic Survey, which was also receiving the observations from the stations and independently determined the origin times and locations of earthquakes. With this service, the JSA, under Macelwane's direction, enlarged its scope and placed itself at the vanguard of seismology.

As dean of graduate studies at Saint Louis University between 1925 and 1933, Macelwane placed a particular emphasis on the need for the science departments to engage in scientific research. On his arrival at Saint Louis, he launched two main research projects in seismology: the study of the propagation of seismic waves in the Earth's interior and the study of regional earthquakes in the central region of the United States. The first produced some of the first tables and curves of travel-times, one of the basic needs for seismological research, while the second continues to the present day, studying the seismic activity of the region. Macelwane also investigated the problem of microseisms, and together with the Colombian Jesuit Jesús Emilio Ramírez, whose doctoral thesis he supervised, tried to apply their observations to determine the path of tropical hurricanes, as we will see below. Another subject he researched was the existence and nature of earthquakes at great depths, a topic very little studied at that time.

In addition to his scientific work, Macelwane also played an important role as a professor and educator in the wider context of Jesuit higher education in the United States. At that time, the principal aim of the Jesuit universities established from the nineteenth century onward was to educate Catholic youth, with research being less of a priority. Macelwane played an important role in changing this situation, moving the emphasis from teaching to include research. Between 1931 and 1932, he chaired a commission established to study the quality of work being done in Jesuit universities in the United States and recommended that greater effort be made to engage in academic research. Macelwane published the first textbook of seismology in North America, *Introduction to Theoretical Seismology* (1936), in two volumes: the first, *Geodynamics*, was written by himself; the second was *Seismometry*, on seismological instrumentation, by Frederick Sohon (1894–1972), director of the Seismological Observatory of Georgetown, and devoted to the theory of the seismograph. Macelwane was also interested in the practical applications of geophysics to prospecting mineral resources and in the development of geophysical engineering. He was particularly interested in the use of seismological methods to detect oil-bearing structures and collaborated with the oil industry. As well as taking an active part in many scientific commissions and congresses, he was also president of various committees on seismology and physics of the Earth of the United States National Research Council. He was appointed president of the Seismological Society of America in 1928 and in 1953 of the American Geophysical Union, the most prestigious scientific societies of seismology and geophysics in the United States. He was also president of the seismological section of the National Science Foundation and had good relations with the research offices of the US Defense Department and Air Force. In 1944, Macelwane was elected a member of the National Academy of Science, the only Jesuit to have received this honor, and in 1962 the American Geophysical Union established a medal in his honor to be awarded each year for special contributions to geophysical sciences by young scientists.

After Macelwane's death, the service provided by the JSA continued until 1963, when it became clear that other institutions and agencies, such as the National Earthquake Information System (NEIS) in the United States and the UK's International Seismological Center (ISC), were doing the same research but with greater means. Most JSA stations updated their instrumentation and continued working until about 1980, and some even continue in operation up to the present day, but with practically no Jesuit presence.

Several Jesuits involved in the JSA project made pioneering contributions to seismological research. Joseph Lynch (1894–1987), for instance, who was director of the Fordham seismological station (first as a young professor between 1920 and 1923 and then for almost fifty years from 1928 to 1977), carried out meticulous observations and published a wide range of scientific articles. In 1946, Lynch traveled to the Dominican Republic at the invitation of its government to study the powerful earthquake that had caused severe damage in the capital Santo Domingo and helped to install the country's first seismographic station; a young Dominican engineer was trained in Fordham to be placed in charge of the station.[73]

Daniel Linehan (1904–87) was director of the Weston Seismological Observatory from 1950 to 1974, which was part of the Jesuit university of Boston College.[74] He initially studied physics at Boston College before going on to study geology at Harvard University. In 1948, Linehan returned as professor

73   Lynch published an interesting account of his long life as a Jesuit seismologist. See Joseph Lynch, S.J., *Watching Our Trembling Earth for 50 Years* (New York: Dodd, Mead and Co., 1970).

74   John F. Devane and James P. McCaffrey, "Reverend Daniel Linehan (1904–1987)," *Bulletin of the Seismological Society of America* 78 (1988): 1618–20.

to Boston College, where he founded the Department of Geophysics, which he directed until 1963, a department that trained several Jesuits who would later go on to work at other observatories. He enlarged the observatory between 1955 and 1980 with a network initially consisting of six stations that was later enlarged to thirty stations, covering the northeast of the United States. In 1940, he also established the first seismographic station in Jamaica, which continued in operation until 1975. Weston Seismographic Observatory continues today as a research institution of Boston College. Linehan's research focused on the applied aspects of seismology, and he made contacts with oil exploration companies that provided him with seismic prospecting equipment that he used in many of his studies of the shallow structure of the Earth's crust. With Lynch's help, Linehan used this technique to carry out archaeological studies under Saint Peter's Basilica in Rome in 1951, the results of which helped in the discovery of a tomb identified as that of Saint Peter. Lynch and Linehan served as presidents of the JSA from 1957 to 1970 and 1970 to 1986 respectively.

Linehan was also famous as an explorer, making many scientific journeys to regions of Africa, Asia, South America, the Arctic, and Antarctica, many of them on UNESCO missions. In 1954, for example, he participated in an expedition to the Arctic that had several scientific aims, including determining the location of the North Magnetic Pole. After a long and hazardous journey in a small ship, Linehan located the North Pole near the Prince of Wales Island, to the north of Canada. Linehan also took part in three expeditions to the Antarctic with the US Navy between 1955 and 1958, where, among other geophysical observations, he determined the thickness of the ice layer by using seismic exploration methods. He was the first priest to celebrate Mass on both the North and South Poles and also the first to perform a baptism in the Antarctic. Linehan was an enthusiastic professor of geophysics, much loved by his students, a prolific writer of scientific and popular science articles, and a frequent speaker at scientific conferences on geology and seismology. He was also frequently invited to speak on subjects related to his expeditions, especially those to the Antarctic. The last Jesuit director of Weston Observatory from 1973 to 1993 was geologist James Skehan (1923–).

Two other Jesuits took also part in expeditions to the Antarctic during the International Geophysical Year (1957–58). Edward Bradley (1923–96), who was director of the seismological station at Xavier University (Cincinnati) and the chief seismologist at the Ellsworth Station (Antarctic), determined the topography of a large area under the ice sheet via the use of seismic exploration methods. A mountain discovered under the ice was named after him as Bradley Nunatak. Similarly, Henry Birkenhauer (1922–2011), director of the station at John Carroll University (Cleveland), also carried out seismic studies in the Antarctic for a period of eighteen months.

In 1960, after receiving his doctorate in geophysics from the University of California, William Stauder (1922–2002) returned to Saint Louis University to take charge of the Department of Geophysics and the seismological stations that Macelwane had established.[75] Stauder gave a strong impetus to seismological studies in Saint Louis University and renewed and increased the number of seismological stations in the surrounding region. He was the last president of the JSA, serving in this capacity from 1987 to 2002. His studies to determine the mechanism of earthquakes from analysis of seismic waves are recognized as having made an important contribution to seismological research. In the 1960s, the theory of plate tectonics was developed, an important landmark in Earth sciences with a new understanding of the dynamics of the Earth, and Stauder contributed to the establishment of this theory; with his study of the mechanism of earthquakes at subduction zones (zones of collision between plates), he helped to clarify how the process took place at plate boundaries. In the 1970s, he turned his attention to the seismic zone at the center of the United States, where four large earthquakes had shaken the region between 1811 and 1812, destroying the town of New Madrid. Stauder installed a network of seismological stations to carry out detailed studies of the small earthquakes that were still taking place in that area. As a result of these studies, which he carried out together with Otto Nuttli (1927–88), he discovered a complex network of faults that cross the region and that had the potential to cause another large earthquake. Besides his scientific work, Stauder was also dean of the graduate school of Saint Louis University and held various offices in scientific societies and commissions.

Among other directors of the stations comprising the JSA network, mention should be made of Louis J. Eisele (1912–88), director for forty-eight years from 1940 to 1988 of the station of Spring Hill, Alabama, one of the few seismological stations in the southern United States. Under his leadership, the station's instruments were regularly updated, and in 1962 he supervised the installation and management of the instruments of the WWSSN station (a worldwide network of seismographic stations funded by the US government). From 1953 to 1970, he sent the information gained from the station to the Associated Press office in Mobile, which was then distributed to the press—and was thus popularly known as "Father Earthquake."

---

75   Brian J. Mitchell and Agustín Udías, "William Stauder (1922–2002)," *Seismological Research Letters* 74 (2003): 373.

The Jesuits also established a seismological station at the Collège Jean de Brébeuf in Montreal, Canada. The station was established after Maurice Buist (1902–86), a professor of science at the college, visited Weston Observatory in 1952, and Linehan encouraged him to install a seismological station in Montreal. Buist installed the first seismographs in the same year, and three years later he built an underground vault where he installed new instruments provided by the Dominion Observatory in Ottawa. Buist conceived the idea of transforming the observatory into a geophysical research center. Gherzi, who, as we saw above, had moved there from China after the closing of Zikawei Observatory, contributed to this idea with his studies of the ionosphere and radiometeorology. In 1960, Buist sent three young Jesuits to be trained for this purpose, but the project did not last long—after their studies, the young Jesuits opted to teach science at state universities, and though the observatory continued its work in seismology, this was solely confined to an observational level. However, the arrival of Gouin in 1975, after his expulsion from Ethiopia, strengthened the observatory's seismological research. In 1979, Gouin published the first catalog of earthquakes in Ethiopia (*Earthquake History of Ethiopia and the Horn of Africa*) and in 2001 a catalog of earthquakes that had taken place in Quebec (*Historical Earthquakes Felt in Quebec from 1534 to March 1925*). Buist, the first and last Jesuit director of the observatory, retired in 1983; the observatory continues today as part of the college, but without any Jesuit members of staff. With no replacements, the Jesuits' active work in seismology in North America, which began with Odenbach in 1908, had largely come to an end by 2000.

9.3   Jesuit Seismologists in Central and South America

Severe earthquakes are a frequent occurrence on the west coast of Central and South America. The Jesuit naturalists of the Old Society, such as José de Acosta (1540–1600) and José Sánchez Labrador (1717–98), had described the occurrence of several significant earthquakes in this region in their natural histories, and the Jesuits of the restored Society in Central and South America also became involved in the study of earthquakes, installing the first seismographs in the observatory of the college of Puebla (Mexico) in 1877. Gustavo Heredia (1869–1926) designed and built the instruments, which consisted of vertical pendulums that recorded horizontal motion on smoked glass circular plates. Heredia was able to record earthquakes in Mexico with these instruments and eventually published his observations for the period 1877 to 1906. These were in fact the first seismographs to be installed on the American continent, a few years before those of California. The seismological station was in operation until 1914 when the observatory and the college were closed. In 1906, the Jesuits installed two Bosch-Omori seismographs in Belén Observatory, the first in the Caribbean region, and they remained in operation until 1930. In 1880, Viñes carried out a survey and published a report of the damage caused by an earthquake in the western part of Cuba. In 1920, the Jesuits also installed a seismological station in Sucre, Bolivia, with horizontal and vertical components, instruments they themselves constructed. However, they were in operation for only a few years; though the instruments were reinstalled in 1945, they were ultimately destroyed in the Bolivian earthquake of 1948. In 1949, German Saa (1909–70) installed two Bosch-Omori seismographs at the Colegio de San Luis, Antofagasta (Chile), and in 1957 he added a Wilson-Lamisson seismograph; the station was eventually closed in 1963 due to the establishment of a nearby seismographic station of the WWSSN network. Saa also intended to establish a seismographic station in Arica in the north of Chile, but this plan never came to fruition.

The Observatory of San Calixto in La Paz (Bolivia), where work in astronomy and meteorology has already been mentioned, established a seismological station in 1912 that acquired a singular importance in South America.[76] The foundation of this station came about following a resolution adopted at the Manchester meeting of the International Seismological Association in 1911, which asked the Jesuits to install a station somewhere in the center of South America. Descotes responded to the resolution by installing the first seismographs in La Paz in 1912 and 1913, instruments he built himself on the basis of Navarro-Neumann's design in the Cartuja Observatory, where he had spent some time training in seismology. In 1926, the station acquired three component Galitzin-Wilip electromagnetic seismographs, the most modern at the time. Under Descotes's guidance, with his very careful and precise observations, the observatory soon came to be regarded as one of the best in South America. In their influential *Seismicity of the Earth* (1949), Charles F. Richter (1900–85) and Beno Gutenberg (1889–1960), two famous Californian seismologists, noted:

La Paz (observatory) at once became and still remains as the most important single seismological station of the world. This is a consequence of its isolated location, the sensitive instruments and the great care with which

---

76   Ramón Cabré, "75 años en la vanguardia de la sismología," *Revista geofísica* 29 (1988): 4–56; Robert R. Coenraads, "The San Calixto Observatory in La Paz, Bolivia: Eighty Years of Operation; Director Dr. Lawrence A. Drake, S.J.," *Journal and Proceedings, Royal Society of New South Wales* 126 (1993): 191–98.

the records were interpreted and the reports issued under the direction of Father Descotes.[77]

As no time signals were broadcast in South America before 1926, Descotes kept exact time on the seismograms thanks to very careful astronomical observations, and over the years the station continued to update its instrumentation.

Ramón Cabré (1922–97) came to the observatory in 1959 and served as its director from 1965 to 1993.[78] Under his direction, the observatory became an important research center in seismology. In 1960, for instance, it participated in an international project to study the structure of the Earth's crust in the Bolivian high plateau (*altiplano*), while in 1962 the station became one of the WWSSN stations, with the best seismographs available. In 1966, Luis Fernández, after obtaining his doctorate at Saint Louis University, came to the observatory and gave a great impulse to seismological research and to the training of young researchers. He left the observatory in 1970 (and also later left the Jesuit order). Under Cabré's direction, the observatory participated in several international seismological research projects. In 1972, a new seismological station was installed in an excavated tunnel in the Zongo Valley with modern equipment provided by the US Geological Survey, one of twelve stations of that type installed worldwide. In the following year, the station became what was called an "Adapted Seismic Research Observatory." Cabré's prestige in South America, which stemmed from his work in seismology, led to him being appointed the first director of the Regional Seismological Center for South America (Centro Regional de Sismología para América del Sur, CERESIS) in 1966, a center depending on UNESCO for seismological cooperation between the South American countries. In recognition of his scientific work, the Bolivian government awarded Cabré the order of Condor de los Andes in 1993, the highest Bolivian award.

Lawrence A. Drake (1931–2007), who had been director of the Riverview Observatory in Australia since 1965, took over the direction of the La Paz observatory after Cabré's retirement in 1992.[79] In La Paz, Drake applied his research on surface waves to determine the complex structure of the high plateau of the Andes; he continued to update the station's instruments and engaged in several research projects on the seismicity and structure of the Earth's crust in the

77 Beno Gutenberg and Charles F. Richter, *Seismicity of the Earth and Associated Phenomena* (Princeton: Princeton University Press, 1949), 6.

78 Richard W. Ott, "Rev. Ramón Cabré Roigé, S.J. May 2, 1922–February 24, 1997," *Seismological Research Letters* 68 (1997): 901.

79 Agustín Udías, "Lawrence Drake, S.J. (1931–2007)," *Seismological Research Letters* 78 (2007): 420.

Bolivian region. He retired as director in 2001 due to poor health and returned to Australia. He was the last Jesuit director of the observatory, which continues to operate today but without Jesuit personnel.

In 1923, Simón Sarasola installed two seismographs at the observatory of San Bartolomé (Bogotá), the first in Colombia. Sarasola resigned from his post as director of the Colombian national meteorological service in order to collaborate with Jesús Emilio Ramírez in the establishment of a new Jesuit observatory devoted to seismology and meteorology. The new observatory was inaugurated in 1941 with the name Instituto Geofísico de los Andes Colombianos (Geophysical Institute of the Colombian Andes).

Sarasola returned to Cuba in 1943, and Ramírez continued as sole director. Under Ramírez's direction, the institute became one of the most important centers of seismological research in South America.[80] Ramírez had been trained in seismology at Saint Louis University where he received his doctorate under Macelwane's supervision with a thesis on the study of microseisms. The seismological station of the institute was located in a tunnel excavated into the rock where modern seismographs were installed. The institute moved to a new building in the campus of the Universidad Javeriana, Bogotá, in 1957, and in 1960 the institute became part of the university when Ramírez was appointed its rector. A WWSSN station was installed in 1962, and in 1975 it became part of the Seismic Research Observatory (SRO) network, a very select worldwide network of only thirteen stations established by the US Geological Survey, with modern seismographs installed in wells at a depth of one hundred meters. The choice of Bogotá as a station in these two networks indicates the prestige gained by Ramírez and the institute in seismological research. For a long time, the institute was in charge of the Seismological Network of Colombia and had its own network of five seismographic stations, along with seven other stations with strong-motion instruments for the observation of motions from near earthquakes.

Ramírez played an important role in scientific education and research in Colombia and South America. From 1947 to 1960, he was professor of geophysics at the Universidad Nacional of Bogotá. In 1967, with Luis G. Durán, he published *Nociones de prospección geofísica* (Elements of geophysical prospecting) a textbook of applied geophysics, the first in South America, and in 1969 the *Historia de los terremotos en Colombia* (History of earthquakes in Colombia), the most complete work on earthquakes in Colombia. Ramírez also published a

---

80   Jesús Emilio Ramírez, *Historia del Instituto Geofísico al conmemorar sus 35 años* (*Universidad Javeriana*) *I. Antecedentes y fundación; II. Los años de la guerra 1944–1945; III. Consolidación del Instituto Geofísico* (Bogotá: Editora Guadalupe, 1977).

large number of scientific articles on the seismicity and tectonics of Colombia. From 1973 to 1979, he organized the Nariño Project, an international project for the study of the seismicity and the structure of the Earth's crust in the Colombian region. At that time, this was one of the most important projects in Earth science to be carried out in South America. Ramírez was also highly active in international seismological circles, serving as president of CERESIS and from 1958 to 1966 as president of the Colombian Academy of Science. In 1971, the Colombian government awarded him the Orden al Mérito "Julio Garavito" in recognition of his work.

After Ramírez's death, the institute was directed by Goberna, who had been expelled from Cuba by the Castro government. The last Jesuit director retired in 1993, and the institute later became part of the Engineering Department of the university with the new name Instituto Geofísico de la Universidad Javeriana. Goberna's death in 1985 and Drake's retirement in 2001 marked the end of the fruitful and long period of Jesuit involvement in seismological research in South America.

9.4   Seismological Stations in Asia, Africa, and Oceania

In 1868, the Manila Observatory installed two homemade seismographs, the first ever installed by Jesuits in Asia, which Faura used to record the severe Manila earthquake of July 18, 1880; the seismograms were the first of an earthquake in the Philippines. John Milne (1850–1913), one of the founders of seismological science, who worked for a period in Japan, recognized the scientific value of the Jesuits' work in Manila and began a correspondence with Faura. In 1881, the observatory installed new seismographs, made in Italy, the best instruments at that time. In 1890, Miguel Saderra-Masó (1865–1939) arrived in Manila with new instruments and reorganized the observatory's seismological section. In 1895, he published the first work on seismology in the Philippines, *La sismología en Filipinas* (Seismology in the Philippines), and later the first catalog of Philippine earthquakes, *Catalogue of Violent and Destructive Earthquakes in the Philippines, 1599–1909* (1910). These two Jesuit works mark the beginning of seismological studies in the Philippines. Saderra-Masó was the first to relate the earthquakes in Philippines with the tectonic accidents in the islands. He installed new instruments in Manila, among them, in 1930, the newly developed Galitzin-Wilip electromagnetic seismographs, and he also established new seismographic stations in Baguio, Butuan, Ambulong, and the island of Guam.

Repetti arrived in Manila in 1928. His doctoral thesis, obtained in St. Louis University under Macelwane's supervision, analyzed the propagation of seismic waves and was one of the first studies to determine the internal structure of the Earth by seismological means. Repetti took charge of the observatory's seismological section and in 1946 published a new catalog of earthquakes in the Philippines.

After the Second World War, the observatory ceased to work on meteorology, and seismology became one of its main research areas, with new seismographs being installed in Manila, Baguio, and Davos. In 1962, the WWSSN program installed two of its stations in Baguio and Davos. The seismographic stations are still in operation with updated instrumentation, but Jesuits are no longer involved in its seismological work.

The observatory of Zikawei installed its first seismographs, Omori instruments provided by the Japanese government, in 1904, thus becoming the first seismographic station in China. Between 1915 and 1932, the observatory installed Galitzin-Wilip electromagnetic seismographs, the best available at that time. Zikawei's observations were highly valued by seismologists given the lack of other stations in the region. The same was also the case for the seismological observations of the observatory of Tananarive (Madagascar), which installed its first seismographs in 1899, the first to be installed in Africa; the observatory updated the seismological instrumentation with Mainka seismographs in 1927. The Jesuits kept the seismological station in operation until it was transferred to the University of Madagascar in 1969. In 1910, Ksara Observatory (Lebanon) installed the first Mainka seismographs, which in 1937 and 1957 were updated with modern electromagnetic instruments that remained in operation until 1979.

In 1907, Pigot installed a set of seismographs at Riverview Observatory, the first in Australia. In 1941, the observatory acquired new Galitzin type instruments made in Australia. Thomas N. Burke-Gaffney (1893–1958), who worked at the observatory from 1946 and was director from 1952 to 1958, won the esteem of seismologists for the observatory's careful analysis of seismological records. For this reason, in 1962 Riverview was also selected as a station of the WWSSN program. Drake was appointed director of the station in 1972 after obtaining his doctorate at the University of California with a thesis on the study of the propagation of seismic surface waves in heterogeneous media. Drake was also a professor of geophysics in Macquarie University and kept in contact with Berkeley via frequent visits. The Riverview seismographic station continues to operate today, but without Jesuit involvement.

9.5   The Study of Microseisms and Seismological Organizations

Microseisms are continuous waves observed in seismograms, with larger amplitudes for periods of between four and ten seconds, which are generated by atmospheric perturbations on the ocean surface. Ever since the start of

instrumental seismology, these waves have attracted the attention of seismologists. In 1903, for example, seismologists Fusakichi Omori (1868–1923) in Japan
and Emil Wiechert (1861–1928) in Germany were among the first to study this
phenomenon and propose its atmospheric origin. It is interesting to note that
a number of Jesuits paid particular attention to the study of microseisms from
the very early years of the restoration. The first was Algué in Manila, who as
early as 1894 linked the occurrence of microseisms registered in the Philippines
with the activity of cyclones on the sea. He dedicated a chapter to microseisms
in his 1904 book on typhoons in which he suggested that seismological stations
located at the coast could be used to detect the proximity of cyclones by the
observation of microseisms. He also suggested that some microseisms could be
caused by the effect of wind on nearby mountains, an idea that is now known
to be incorrect. Gherzi made microseisms the main object of many of his studies in Zikawei.[81] In 1924, he announced that microseisms could be divided into
four classes according to their origin: those due to cyclones or centers of low
atmospheric pressure, those due to anticyclones or high pressure, those caused
by cold, and those with no known cause. He considered the first type to be
the most important: they are caused by sudden changes of atmospheric pressure in the vortices of cyclones that are transmitted through the water layer
of the oceans to the Earth's crust. He also proposed that, in stations at the sea
coast, the effect of the sea waves on the coast should be taken into account. At
Saint Louis University, Ramírez, under Macelwane's supervision, studied the
nature of short-period microseisms and their generation by low atmospheric
pressure over the oceans. Using the idea that microseisms are caused by storm
centers, Ramírez thought it might be possible to determine the location of
tropical cyclones by using a system of three seismological stations located at
the vertices of a triangle.[82] The US Navy soon became interested in this idea,
which it attempted to use to locate the centers of cyclones in the Pacific and
Atlantic Oceans. However, the system proved highly inefficient in practice and
was eventually abandoned.

Gherzi's interest in microseisms led him to suggest to the Pontifical Academy
of Sciences the organization of a study week devoted to microseisms. Francesco
Vercelli (1883–1952), director of the Geophysical Observatory of Trieste, and
a member of the academy, supported the idea and acted as chairman of the

81   Ernesto Gherzi, "Étude sur les microséismes," *Observatoire de Zikawei: Notes sismologiques*
      5 (1924): 1–16.

82   Jesús Emilio Ramírez, "An Experimental Investigation on the Nature and Origin of Mi-
      croseisms at St. Louis, Missouri," *Bulletin of the Seismological Society of America* 30 (1940):
      34–84; 139–78.

study week. The study week took place at the offices of the academy in the Vatican from November 19 to 26, 1951.[83] Sixteen eminent seismologists took part: among them were Maurice Ewing (1906–74) of Wood Hole Oceanographic Institute; Inge Lehmann (1888–1993) of the Geodesic Institute of Copenhagen, discoverer of the inner core of the Earth; Jean Pierre Rothé (1906–91) of the Institut de Physique du Globe of Strasbourg; and Robert Stoneley (1894–1976) of Cambridge University and president of the International Association of Seismology. Three Jesuits were also present: Gherzi, Macelwane, and Due-Rojo. The study week sought to analyze the problems involved in identifying the nature and origin of microseisms. On the first day, Pope Pius XII gave an important talk on the relationship between science and theology.

A year later, the United States Naval Research and the Geophysical Research Directorate of the US Air Force organized a similar symposium on microseisms. Three Jesuits, Macelwane, Lynch, and Ramírez, were among the thirty-one specialists taking part. However, the interest in microseisms shown by Jesuits was limited to those of this period, and there was no continuation of Jesuits working on this problem. Indeed, Jesuit interest in seismology decreased drastically after 1980—from that year onward, most Jesuit seismographic stations either closed or handed over their direction to laypeople.

The Jesuit contribution to seismology coincided with the beginning of the twentieth century and with the establishment of the first national and international seismological organizations. A number of Jesuits took part in these organizations. In 1901, the first International Conference of Seismology took place in Strasbourg, when a decision was made to establish the International Association of Seismology (IAS), which began operating in 1905; Cirera, the first director of the Ebro Observatory, participated as one of the two Spanish delegates in the second International Conference of Seismology in 1903. In 1907, three Jesuits, Berloty (Lebanon), Sánchez Navarro-Neumann (Spain), and Stein (Vatican), were present at the first general assembly of the IAS in The Hague, which in 1922 became part of the International Union of Geodesy and Geophysics (IUGG). In 1936, the IAS created a commission that met in Edinburgh to supervise the International Seismological Summary, an agency that published seismological data. Macelwane was one of its five members. In 1939, a meeting of the IAS took place in Washington, the first in North America, and nine Jesuits, including Macelwane, Lynch, and Linehan, were among the seventy-four American participants. Macelwane presented a report at the meeting on the problem of microseisms, a problem on which some Jesuits

---

83   *Semaine d'étude sur le problème des microséismes*, Scripta Varia 12 (Vatican City: Pontificia
       Academia Scientiarum, 1952).

were working at that time, as we have seen. In 1951, the IAS changed its name and became the International Association of Seismology and Physics of the Earth's Interior (IASPEI). The IASPEI meets every two years, and Jesuit seismologists attended these meetings for as long as they continued to manage seismographic stations.

Some Jesuit directors of seismographic stations also occupied important positions in national seismological commissions and other organizations. Cabré and Ramírez, for example, were presidents of CERESIS, and Ramírez was also vice president of the Seismological Committee of the Pan-American Institute of Geography and History. American Jesuit seismologists had a special relationship with the Seismological Society of America (SSA). Ricard, director of the Santa Clara Observatory, was one of the thirteen founding members of this society and was elected to the first directive committee. Macelwane was a member of the committee between 1925 and 1956 and in 1928 was elected president of the society. Stauder, who was a member of the committee between 1962 and 1967, was appointed its president in 1966. Several Jesuits also served as presidents of the SSA's Eastern Section, including Macelwane, who was appointed in 1926, Lynch in 1930, Linehan in 1954, Birkenhauer in 1956, and Stauder in 1963. The JSA held several joint meetings with the SSA's Eastern Section. At a meeting in Saint Louis in 1934, for example, nine of the twenty-two participants were Jesuits, and in 1948, in Cleveland, Jesuits comprised eight of the twenty-nine participants. In 1961, at the meeting in Cincinnati, eight of the eighty participants were Jesuits. In 1991, the SSA Eastern Section established a JSA medal and award in recognition of the Jesuits' pioneering work in seismology, which is bestowed each year on seismologists distinguished for their work in observational seismology. In 1997, Stauder became the only Jesuit to receive this award.

10   The Tradition in Mathematics

The Jesuits' contribution to modern science in the Old Society was the work of professors of mathematics such as Clavius, the founder of the tradition, Grégoire de Saint Vincent (1584–1667), Giovanni Saccheri (1667–1733), and André Tacquet (1612–60), to mention only a few. After the Society's restoration, this tradition was continued by the Jesuits of the New Society, a number of whom excelled as professors of mathematics. Most worked at Jesuit universities and colleges, where they were mainly involved in teaching and writing textbooks at elementary and some at higher levels. A few who made some more important contributions are discussed in further detail below.

One such figure was the Italian Joseph Bayma (1816–92), who worked as a professor of mathematics, initially in Stonyhurst College, where he moved in 1860, and later at Santa Clara University, California, where he published several textbooks on calculus and analytical geometry. He proposed a mathematical theory on molecular mechanics that resembled that proposed by Roger Boscovich. The Irishman James Cullen (1867–1933), who worked as a professor of mathematics at Mount St. Mary's College, Derbyshire, from 1905, focused in particular on number theory, proposing what are now called Cullen numbers. Hagen, whom we earlier encountered as director of the Georgetown and Vatican observatories, also published an extensive treatise on higher mathematics (*Synopsis der höheren Mathematik*). Similarly, Maurice Proton (1872–1942), a graduate of the École Polytechnique who received a doctorate for his thesis on number theory, taught mathematics at several places including the Catholic Faculties of Paris and Lille and the Jesuit schools of Saint-Geneviève and Saint François-Xavier. Although he also published works on number theory, his main contribution was his pioneering work on the application of mathematics to economics. In Spain, Enrique de Rafael (1885–1955), who joined the Jesuits after having already gained a doctorate in mathematics, initially taught in India at Xavier's College (Mumbai) and then at ICAI (Madrid), where he developed the theory of so-called saturated numbers. Alberto Dou (1915–2009) was professor in Madrid at the School of Civil Engineering and Universidad Complutense and worked in particular on the theory of ordinary and partial differential equations. These are just a few examples of the many modern Jesuit professors of mathematics at Jesuit colleges and universities and at state institutions who made significant contributions to this science.

11   The New Naturalists and Biologists

During the seventeenth and eighteenth centuries, many Jesuit naturalists—men such as José de Acosta and Bernabé Cobo (1580–1657) in South America and João Loureiro (1717–81) and Georg Kamel (1665–1706) in Asia—carried out important work in botany and biology, producing the first studies of the flora and fauna of the Americas and Asia and making them known in Europe, often for the very first time. This tradition continued in the modern Society of Jesus thanks to the Jesuits teaching in universities, faculties of philosophy, and secondary schools.

Biology itself is a relatively modern science. In the mid-eighteenth century, Carl von Linné (Linnaeus) (1707–78) used biological taxonomy to establish a basis for the classification of living beings, and it was only at the beginning of the nineteenth century that the term "biology" was used to refer to this science by Gottfried Reinhold Treviranus (1776–1837) and Jean-Baptiste Lamarck (1744–1829). As in the case of meteorology and seismology, Jesuits contributed to the development of biology as a science, especially in the fields of botany and entomology, mainly in the early part of the twentieth century.

The first group of Jesuit biologists was based at the German faculties of philosophy in Valkenburg, Pullach, and Frankfurt. The most notable figure of the period was the Austrian Erich Wasmann (1859–1931), who studied in Vienna and Prague and devoted his life to the study of ants, termites, and their parasites.[84] He was on the staff of the Jesuit journal *Stimmen der Zeit* (Voices of the time) in which he published many of his articles on biological subjects, including a large number of studies on ants and termites. He also published two books in which he describes their system of communication and social organization. During the course of his career, he collected and classified some one thousand species of ants, two hundred species of termites, and two thousand species of myrmecophiles, 933 of which were new species. Wasmann was an early defender among Catholic authors of the theory of evolution, though in a moderate version that rejected natural selection as the only mechanism involved as well as some of the applications of the theory to the descent of man. His *Die moderne Biologie und die Entwicklungstheorie* (Modern biology and the theory of evolution [1906]) had a wide circulation, running to three editions and several translations. Wasmann claimed that the theory of evolution was not opposed to Christian faith and that divine providence guides the evolutionary process according to the internal laws of nature, but opposition from conservative sections in the church forced him to stop publishing further editions of his book. He took part in a long debate with Ernst Haeckel (1834–1919), whose philosophical monism he rejected in two series of conferences held in Berlin in 1905 and 1907, instead proposing a Christian monism. The University of Freiburg (Switzerland) awarded him an honorary doctorate in 1921. The Jesuit University of San Francisco published a journal under his name (*Wasmann Collector* [1936–49]) and *Wasmann Journal of Biology* (1950–94). One of Wasmann's followers, Karl Frank (1875–50), combined his work as a biologist with his role as professor of natural philosophy in the Jesuit Faculty of

---

84   Charles J. Wideman, *Erich Wasmann, S.J. 1859–1931* (San Francisco: Wasmann Biological Society, 1942); Christian Kummer, "Forschung und Weltanschauung: die Rolle der deutschen Jesuiten in der Biologie," *Sammelblatt des Historischen Vereins Ingolstadt* 109 (2000): 105–20. Abigail J. Lustig, "Erich Wasmann, Ernst Haeckel and the Limits of Science," *Theory in Biosciences* 121 (2002): 252–59.

Philosophy in Pullach (Bavaria). Another collaborator of Wasmann in *Stimmen der Zeit* was Felix Rüschkamp (1885–1957), who studied in Bonn, eventually receiving a doctorate for his thesis on beetles and becoming a professor in the Jesuit faculty of philosophy at Frankfurt; however, he ultimately resigned as professor after being forbidden from writing about the problem of the origin of man. Adolf Haas (1914–82) succeeded Frank in Pullach; his doctoral thesis was on the flight of bumblebees, and he worked with ethologist Konrad Lorenz (1903–89) on the behavior of certain fish and on the concept of instinct in animals. Haas contributed to the spread of Teilhard de Chardin's thought in Germany and to the development of the theory of divine creation through evolution.

There was also another group of Jesuit biologists based in Spain. In 1897, Longino Navás (1858–1938), a professor in the Jesuit school in Zaragoza, began working as an entomologist specializing in Neuroptera, classifying up to 3,316 new forms (388 genera, 2,684 species, and 244 varieties).[85] A tireless worker, he published 588 articles in scientific journals and proceedings of congresses between 1898 and 1935. One of his most important publications was *Entomologia de Catalunya: Neuròpters* (Entomology of Catalunya: Neuroptera, 3 vols. [1923, 1926, 1934]). In 1915, Pelegrín Franganillo (1873–1955) began his research with a study of spiders in Galicia and Asturias in northern Spain, which he eventually published as *Las arañas: Manual de aracnología* (Spiders: Handbook on arachnology [1917]). In 1918, he moved to the Colegio de Belén in Havana where he continued his work on spiders and published *Arácnidos de Cuba* (Spiders in Cuba [1923]). Besides his studies of the climate, Merino also collected and classified a large herbarium of the flora of Galicia (northwest Spain) with more than twenty thousand specimens. Jaime Pujiula (1869–1958), the founder of the Biological Institute of Sarriá, Barcelona, who had studied in Innsbruck and Trieste before obtaining a doctoral degree in Vienna, was a pioneer in embryological, histological, and cytological research in Spain. He was a member of the Academy of Medicine of Barcelona and published *Citología* (Cytology [1928]).

The two Portuguese Jesuit colleges of Campolide (Lisbon) and São Fiel (Louriçal do Campo) played a particularly important role in the development of biology as a science. Joaquim da Silva Tavares (1866–1931), Cândido Azevedo Mendes (1874–1943), and Carlos Zimmermann (1871–1950), all teachers at the Jesuit college of São Fiel, founded the journal *Brotéria* (named after the Portuguese botanist Félix de Avelar Brotéro [1744–1828]), which was originally

---

85   Juan Jesús Bastero Monserrat, *Longino Navás, científico jesuita* (Zaragoza: Universidad de Zaragoza, 1989).

dedicated (between 1902 and 2002) to the natural sciences, and especially botany, zoology, and genetics—it continues to be published today as a cultural journal.[86] Tavares, an entomologist and member of the Academy of Science of Lisbon, worked on the cecidia and zoocecidia of Portugal, Brazil, Argentina, and Madeira. Mendes's research focused on the study of butterflies. In the college of Campolide, António Oliveira Pinto (1868–1933) organized an institute of natural sciences with six sections of physics, chemistry, mineralogy, geology, botany, and zoology. He was a professor of physics and an early specialist in radioactivity. Camillo Torrent (1875–1961), who worked at the same college, formed a very complete collection of Myxomycetes. From 1962 to 1979, the director of *Brotéria* was Luis Jorge Archer (1926–2011), a professor of molecular genetics at the Universidade Nova in Lisbon and a research fellow at the Gulbenkian Institute of Sciences. He published a long list of publications, and the Portuguese Society of Genetics has established an award in his honor.[87] Other noteworthy Jesuit naturalists include Xavier Ferenc Speiser (1854–1933), who taught at the school at Kalocsa, Hungary, and was an entomologist. He studied 9,500 species of coleopterans, most of them from Africa, which László Menyhart, also a professor at Kalocsa, collected during his journeys to the Zambezi region.

In South America, Balduino Rambo (1905–61), professor at the Colegio Anchieta in São Leopoldo (Brazil), followed in the tradition of naturalists of the Old Society by founding the Museum of Natural Sciences of Rio Grande do Sul and organizing the Herbario Anchieta, which contained more than sixty thousand plants. Luis Sodiro (1836–1909), an Italian, worked in Ecuador from 1870 to 1909 describing a total of 281 taxa of Ecuadorian Araceae, largely of the *Anthurium* species. He was the first botanist specializing in aroids to be considered a field botanist and spent most of his career in the Ecuadorian tropics. He founded the botanic Garden of Quito and collected a large herbarium, which is currently located at the University of Quito. In Colombia, Antonio Uribe (1900–81), a professor at the Universidad Nacional, was an important contributor to the Herbario Nacional and author of *Pasifloráceas y Bignoniáceas de la expedición botánica del Nuevo Reino* (Passifloraceae and Begoniaceae of the botanic expedition of the New Kingdom [1956]). The English botanist Gerald Wilson Browne (1911–75) studied the flora of Kanuku in Guyana.

---

86   Francisco Malta Romeiras, "The Journal *Brotéria* (1902–2002): Jesuit Science in the 20th Century," *Journal of History of Science and Technology* 6 (2012): 100–9; Malta Romeiras, "A ciência da Companhia de Jesus nos séculos XIX e XX em Portugal," *Brotéria* 179 (2014): 429–54.

87   Francisco Malta Romeiras, "The Emergence of Molecular Genetics in Portugal: The Enterprise of Luís Archer," *Archivum historicum Societatis Iesu* 82 (2013): 501–12.

The Jesuits also carried out important biological research in China. The French Jesuit Pierre Marie Heude (1836–1902), for instance, who traveled to China in 1868, devoted all his time and energy to the studies of natural history, traveling widely in China and other parts of eastern Asia. His research initially centered on mollusks, before moving on to the study of mammals and birds in China; he set up a museum of natural history at Xujiahui (Zikawei) in 1868, the first of its kind in China. The French naturalist Émile Licent (1876–1952) also worked in China, spending more than twenty-five years working in the region of Tianjin. This short list provides merely some of the examples of the many Jesuit professors of natural sciences who besides their teaching duties in colleges and universities dedicated time and a great deal of effort to the study of the fauna and flora in many parts of the world.

11.1   *Botany and Ecology in India*

At the beginning of the twentieth century, Jesuits in India began producing an important series of botanical studies in relation to agriculture, environmental studies, and ecology, and much of this work continues to the present day.[88] Moreover, as well as creating and maintaining botanical gardens, Jesuits also promoted reforestation and the preservation of rare, endangered, and threatened plant species, together with other agricultural programs. A Swiss Jesuit, Ethelbert Blatter (1877–1934), who arrived in India in 1903, was among the first to engage in this type of work.[89] He taught at St. Xavier College, Mumbai (Bombay), and later became a member of the Bombay Natural History Society. He studied and published extensively on different botanical subjects such as mangroves, ferns, mosses, grasses, and many other varieties of flora of the Mumbai region. He established a collection of nearly three million plants in the college, which is known today as the Blatter Herbarium. His many publications include *The Palms of British India and Ceylon* (1926), published by Oxford University Press, *Beautiful Flowers of Kashmir* (2 vols. [1927–28]), and, in collaboration with Walter S. Millard (1864–1952), *Some Beautiful Indian Trees* (1937, 1997). The Asian Society of Calcutta acknowledged his work on botany by awarding him the first Bruehl Memorial Award. Another noteworthy professor and successor of Blatter at St. Xavier's College was Hermenegildo Santapau (1903–70), from Tarragona (Spain); he was an accomplished botanist with a doctorate from

---

88   Job Kozhamthadam, "Jesuit Contribution to Science in India," *Vidyajyoti: Journal of Theological Reflection* 68 (2004): 892–916. Leo D'Souza, "Jesuit Contribution to Environmental Protection," 2009; http://www.sjweb.info/sjs/networks/ecology/Indian%20Jesuits%20 Contribution%20to%20Environmental%20Protection.pdf (accessed August 23, 2018).

89   Charles McCann, "Ethelbert Blatter, S.J.," *Journal of the Bombay Natural History Society* 37 (1934): 465–73.

London University and had spent some time at the Royal Botanical Gardens in Kew, England.[90] In 1955, he was appointed president of the Indian Botanical Society, and between 1961 and 1967 he was director of the Botanical Survey of India. He published some 216 scientific articles and reports, as well as several books, including *The Flora of Purandar* (1958) and *The Flora of Saurashtra* (1962). In recognition of his work, the government of India awarded him the Padma Shri Award in 1967.

At St. Joseph College, Tiruchirappalli, Alfred Rapinat (1892–1959) collected a large number of plant specimens and portraits that later became known as the Rapinat Herbarium, thereby continuing Pierre Labarthere's (1831–1904) earlier work on botanical gardens at the same college. Another botanist at St. Joseph College, Joseph Mathen Pallithanam (1915–84), was a tireless worker in the field of taxonomy who studied the South Indian flora. Koyapillil M. Matthew (1930–2004), born in Kerala, succeeded Rapinat and Pallithanam as professor of botany at St. Joseph College. He organized the Rapinat Herbarium and worked with great energy for the study and conservation of the *shola* forests of the subtropical regions of south India. Well known as a taxonomist, he published, among other books, *The Flora of the Palni Hills* (1999) and *The Flora of the Tamilnadu Carnatic* (5 vols. [1983]), and using the results from Pallithanam, *A Pocket Flora of Sirumalai South India* (2001). Matthew organized the Rapinat Herbarium, a collection of more than fifty-four thousand plants collected and cataloged by Jesuits since 1895; he also encouraged the local people to take an interest in their plants and their preservation, developing training programs on environmental conservation techniques. In 2002, the government of India awarded him the Indira Gandhi Paryavaran Puraskar Award in recognition of his environmental studies. In another part of India, Cecil Saldanha (1926–2002), professor of botany at St. Joseph's College, Bangalore, studied the flora and the environment of Karnataka state, research he eventually published in a volume entitled *Flora of Karnataka* (2 vols. [1978]), with descriptions of 1,700 species. Visuvasam S. Manickan (1941–2012) at St. Xavier College (Palayamkottai) was a specialist on the ferns of the west Ghats and Tamil Nadu region. He established a fernery for conservation purposes and a center for biodiversity and biotechnology.

Botany, applied botany, environmental, and ecological studies are taught today in many of the Jesuits' Indian colleges, including those of St. Xavier's (Mumbai), St. Xavier's (Ahmedabad), Loyola (Chennai), St. Joseph's (Tiruchirappalli), St. Aloysius (Mangalore), St. Xavier's (Kolkata), and St. Xavier

---

90   P. V. Bole, "Rev. Fr. Hermenegild Santapau," *Taxon* (*International Association for Plant Taxonomy*) 19 (1970): 576–83.

(Palayamkottai). They promote ecological awareness and conservation activities, together with other environmental projects such as the use of biopesticides, biofertilizers, the preservation of rare and endangered plant species, and the promotion of solar energy.[91]

12   Jesuit Scientists in Non-Jesuit Institutions

Throughout the post-restoration period, the Jesuits who were active in various fields of science were not confined to their own universities and schools, as Jesuit scientists also worked at non-Jesuit universities and research centers, both public and private, and many continue to do so today. This motivation to work in non-Jesuit universities is often driven by the need to participate in large and prestigious scientific institutions where cutting-edge research is produced. Thus, besides often obtaining their degrees at these institutions, some Jesuits chose to remain working at them. For Jesuits more interested in scientific research than teaching, this is in many cases a necessity, and it is likely to become an increasingly common practice in the future.

One of the most notable Jesuits to have worked primarily in non-Jesuit institutions was Teilhard de Chardin,[92] who worked at the Museum of Natural History in Paris, the Geological Survey, and the Museum of Man in Beijing, and in his final years at New York's Wenner-Gren Foundation. His scientific work was especially linked to the study of the fossil origins of man in China, and he participated in the discovery in 1929 of the Peking man (*homo erectus Pekinensis*) at Zhoukoudian (Choukoutien). Teilhard de Chardin, who was highly regarded among the scientific community, made frequent journeys to France and the United States and also carried out field studies in Kashmir, Java, Burma, and South Africa in addition to his work in China. As both a Jesuit priest and a passionate scientist, Teilhard de Chardin was particularly concerned with harmonizing Christian thought with the new worldview generated by modern science of a universe in evolution. In his philosophical and theological work, Teilhard de Chardin produced an original philosophy inspired by a true mystical

---

91   Savarimuthu Ignacimuthu, "The Contributions of South Asian Jesuits to Environmental Work," *Journal of Jesuit Studies* 3, no. 4 (2016): 619–44.

92   There is an extensive amount of literature on Teilhard de Chardin. See, e.g., Claude Cuénot, *Pierre Teilhard de Chardin: Les grands étapes de son évolution* (Paris: Librairie Plon, 1958); Robert Speaight, *Teilhard de Chardin: A Biography* (London: Collins, 1967); Ursula King, *Spirit of Fire: The Life and Vision of Teilhard de Chardin* (New York: Orbis, 1998); Patrice Boudignon, *Pierre Teilhard de Chardin, sa vie, son oeuvre, sa reflexión* (Paris: Éditions du Cerf, 2008).

experience, which he expressed in a large number of essays and two books, *The Human Phenomenon* and *The Divine Milieu*, which he was never allowed to publish during his lifetime.[93] It is customary to see Teilhard de Chardin either as a scientist specializing in geology and paleontology, which put him in contact with the early history of the human, or as a philosopher who elaborated a new evolutionary synthesis of man and the cosmos. However, he was also a Christian mystic who discovered the cosmic Christ in the fibers of matter and as the ultimate end of the evolution of the universe.[94]

Among other Jesuits, Lejay, who was director of the observatory of Zikawei, went on to work at the Laboratoire National de Radioélectricité of the Centre National de la Recherche Scientifique (CNRS) after his return to France. Mayaud also had an official post as maître de recherches at the CNRS. Archer worked at the Universidade Nova in Lisbon and at the Gulbenkian Institute of Sciences. Some Jesuits working at Jesuit observatories were also professors at other centers. Besides being director of Riverview Observatory, for example, Drake was professor at the Macquarie State University. Similarly, Ingram, director of the observatory of Rathfarnham Castle, also taught mathematics at University College Dublin, as we saw earlier. Wulf and a group of German Jesuits managed the new state institution Polytechnic School in Quito, Ecuador, for a time, as we saw above, and Sarasola and Ramírez were briefly in charge of the national meteorological service of Colombia. Macelwane organized studies and research in seismology at the University of California, Berkeley, for a couple of years.

During the 1970s, a few Jesuits worked for the CNRS in France and at the Institut Catholique in Paris and Toulouse, where they primarily carried out biological research. George Ruggieri (1925–1987), a marine biologist, was director from 1976 to 1987 of the New York Aquarium. In the same period, there were Jesuit professors of science at state universities in other countries in Europe, such as the University of Padua (Italy), Warsaw (Poland), and the National University (Ireland).

Conversely, in countries such as the United States and India, where there are many Jesuit colleges and universities, there are only a small number of examples of Jesuit scientists who worked in other institutions. One interesting exception is Timothy Toohig (1928–2001), a nuclear physicist associated with

---

93   Teilhard de Chardin, *Oeuvres*, vols. 1–13 (Paris: Éditions du Seuil, 1955–76).

94   Christopher F. Mooney, *Teilhard de Chardin and the Mystery of Christ* (New York: Doubleday Image Book, 1968); Édith de la Héronnière, *Teilhard de Chardin, une mystique de la traversée* (Paris: Albin Michel, 2003); Gustave Martelet, *Teilhard de Chardin, prophète d'un Christ toujours plus grande* (Brussels: Lessius, 2005).

Boston College. Initially, he carried out his research in Brookhaven National Laboratory between 1965 and 1970, later moving to the Fermi National Accelerator Laboratory (Fermilab), a center dependent on the US Atomic Energy Commission, where he collaborated on the design of the Tevatron, a particle accelerator. From 1988 onward, he worked with the project for the new particle accelerator, the Superconducting Super Colliding (SSC), which was never actually finished. He also served as one of the US representatives in the Conseil Européen pour la Recherche Nucléaire (CERN) in Geneva.

Due to the expansion of vocations in the 1950s and an increase of Jesuits who had studied in state universities, from 1970 to 1985 there was a large number of Jesuit scientists working at state institutions in Spain. Many remained in these institutions as professors after completing their studies, while others held research posts in the Higher Council of Scientific Research (Consejo Superior de Investigaciones Científicas [CSIC]). For those who had studied science, there was the additional factor that the only existing Jesuit higher educational institutions with programs in science in Spain were the IQS in Barcelona and the ICAI in Madrid, both with a marked focus on engineering. The number of Jesuits in state institutions in 1973 reached a high of seventy-five, made up of professors and research specialists working in different fields; twenty-seven of them worked in scientific subjects. In the same year, the provincial of Spain called a meeting in Madrid for Jesuits involved in scientific research, and in 1977 he created a group under the title "University Mission in Non-Jesuit Institutions" (Misión Universitaria en Instituciones que No son de la Compañía de Jesús [MUINSI]). He appointed Julián Rubio (1925–96), a professor of genetics in the University of Oviedo (northern Spain), to be the coordinator of the group; it went on to meet every year thereafter. Alberto Dou, professor at the Universidad Complutense (Madrid), played an important role in the group. An evaluation in 1980 of the apostolic value of this experiment recognized the positive value of having members of the Society of Jesus and of the church in state institutions and in particular in scientific fields. In those years, among the Jesuit scientists occupying posts in state universities, there were eleven top-level professors (*catedráticos*) in the fields of mathematics, theoretical physics, atomic physics, optics, geophysics, physical chemistry, biochemistry, biology, genetics, and paleontology at the universities of Madrid-Complutense, Barcelona, Barcelona-Autónoma, Barcelona-Politécnica, Seville, Oviedo, Málaga, and Cordoba. Superior General Pedro Arrupe (1907–91, in office 1965–83) saw this experience as a way to maintain the Jesuit presence in the university world in both the natural sciences and the humanities. Annual meetings of the MUINSI continued until 1988, when they stopped because no new Jesuits had joined any of the state universities. Hence the main reason the group was disbanded was the sharp decline in the number of new Jesuits in Spain; the number of novices fell from about one hundred per year to fewer than ten. At the same time, the departure from the Society of some young Jesuits dedicated to science influenced the superiors' decision not to continue this kind of work, together with changes in the choice of apostolic priorities, with preference now being given to social and pastoral work and theological and spiritual studies.

13   Recent Developments

At an official level, the Thirty-First General Congregation (1965) was the last to refer explicitly to the work of Jesuits in the natural sciences (decree 29 on scientific research). The decree begins: "Jesuits must greatly value scientific work and especially authentic research and must consider it as one of the most important works of the Society."

Although the decree refers in the first place to the sacred sciences, it also mentions "the mathematical–natural sciences and the technology that is derived from them."[95] The decree justifies such dedication given the mentality of the time, which is so deeply penetrated by science. However, the following Thirty-Second General Congregation (1974) unintentionally had a negative influence on the Jesuits' involvement in scientific research, especially with decree 4 (On the Service of Faith and Promotion of Justice). This decree marked a change in the priority given by Jesuits to the social apostolate, and although not intended to do so, in practice it led to a diminution of the Jesuit involvement in the natural sciences. Nothing was explicitly stated about the intellectual apostolate in the natural sciences, and the general congregations that have taken place since, namely the thirty-third (1983) to the thirty-sixth (2016), barely made any mention of scientific work as part of the Society's intellectual apostolate.

After the Thirty-Second General Congregation in 1976, Superior General Pedro Arrupe put forward a compelling argument in favor of Jesuits' dedicating themselves to the natural sciences in a letter on the intellectual apostolate:

> How can we make the Church present, and keep the necessary personal contacts in a social context of so vital importance as the scientific and technological, without giving to science the importance it deserves? How

---

95   Decreta Congregationis Generalis 31, "De labore et investigatione scientifica," *Acta Romana S.I.* 14 (1961–66): 958–60.

can we make a theological reflection that is intelligible without a profound knowledge of the scientific roots of this mentality?[96]

Hence the superior general continued to view scientific research as an important part of the Jesuits' work. These questions still remain relevant today,
when the Jesuit presence in scientific fields has radically diminished, as we
will see below.

Soon after the restoration, Jesuits working in the natural sciences felt the
need to form associations to help them in their work, distinguishing them to
some extent from other Jesuits more involved in the humanities, theology,
or pastoral work. The first was the American Association of Jesuit Scientists,
established in 1922 in the United States, membership of which was initially
restricted to Jesuits from the northeastern states, but later comprised other
Jesuits from across the United States. In 1954, when Jesuits in science were at
a high, the association had 254 members; it met once a year and published
a quarterly bulletin from 1922 to 1966,[97] which remains an important source
of information about Jesuit scientific activity in the United States during this
period. In 1968, Pedro Arrupe, concerned about the declining involvement
of Jesuits in science, appointed François Russo (1909–98), a philosopher and
historian of science, to be the international coordinator of Jesuits working in
the positive sciences. In 1970, Russo began publishing the *Bulletin de liaison
des jesuites scientifiques* (Bulletin of coordination of Jesuit scientists; six issues
until 1975). Charles Courtoy (1917–2008), professor of physics at Université de
Namur, succeeded Russo in 1976, and in 1982 Alberto Dou became coordinator.
Dou began the publication of a new bulletin in 1985, *Jesuits in Science* (fifteen
issues up to 1999, when it ceased publication).

In Europe, a group named European Jesuit Scientists met for the first time
in 1989 at Aix-en-Provence (France). Meetings continued every two years until
2009, with twenty to thirty participants. However, the meetings stopped when
it was made clear that there was practically no new incorporations of young
Jesuit scientists.

A more recent development associated with the presence of Jesuits in science is the activity of Jesuits on the relationship between science and religion. We have already seen the work carried out on this subject by the Vatican
Observatory in collaboration with the CTNS since 1988, especially by George

---

96   Pedro Arrupe, "El apostolado intelectual en la misión de la Compañía hoy," *Acta
     Romana S.I.* 16 (1973–76): 1015.
97   *Bulletin of the American Association of Jesuit Scientists* (*The Jesuit Science Bulletin*),
     vols. 1–43 (1922–66).

Coyne and William R. Stoeger (1943–2014). The observatory participated in the publication of books on the subject such as *Physics, Philosophy and Theology* and *Scientific Perspectives on Divine Action* and the organization of courses and seminars. In 1968, Robert A. Brungs (1931–2006), a professor of physics at Saint Louis University, co-founded with Dr. John Matschiner (1927–2003) (a biochemist) the Institute for Theological Encounter with Science and Technology (ITEST) and was its director for thirty-eight years. This is one of the oldest institutes engaged in active work on the relationship between science and religion; it publishes a quarterly bulletin and organizes conferences and workshops. Since 2000, the Gregorian University has collaborated with six other catholic universities in the project "Science, Theology and the Ontological Quest" (STOQ) to further the connections between science, philosophy, and theology. Similarly, in 2003, the Universidad Pontificia Comillas (Madrid, Spain) established a chair of science, technology and religion (Ciencia, Tecnología y Religión, CTR) dedicated to this subject with a variety of activities. The first to hold this chair was Javier Leach (1942–2016), a professor of mathematics at the Universidad Complutense (Madrid). Several Jesuit scientists from different fields participate in these projects.

However, as we have seen throughout this paper, from around 1980 onward there has been a downward trend in the number of Jesuits involved in scientific research in both observatories and universities. With the exception of the Vatican Observatory, which, as we have seen, is actually a pontifical institution entrusted to the Jesuits, as well as a few other cases, all the observatories have been closed or entrusted to other hands. Those that are still part of Jesuit institutions are mainly staffed and run by laypeople. As pointed out above, the number of Jesuit science professors in colleges and universities has also diminished radically. The new situation deserves further analysis in greater detail, a subject to which the paper turns below.

The period from 1950 to 1980 saw the highest number of Jesuit scientists in modern times. A fairly complete overview can be gained from a survey compiled by the coordinators of Jesuits in science, Russo and Dou, between 1970 and 1980.[98] The survey covered professors in Jesuit universities and colleges, Jesuits working in observatories, some (though not all) professors in secondary schools, and those working in non-Jesuit institutions. In the four areas— mathematics, physics, chemistry, and biology—and excluding engineering and medical sciences, the total number came to 547: mathematics 152, physics 177,

---

98   Information about the number of Jesuit professors of science in the different scientific
      fields in 1975 obtained from the survey of the International Coordinator of Jesuits Work-
      ing in the Positive Sciences, made by François Russo and Alberto Dou between 1975 and
      1985 (Biblioteca Comunidad ICAI, Madrid).

chemistry eighty-nine, and biology 129. Thus most worked in physics, which included astronomy, geophysics, and meteorology, followed by mathematics and biology. The lowest number was in chemistry. The two countries with the highest numbers were the United States (181) and India (136). In Europe, there were 127 (forty-four of whom worked in Spain). There were also fifty-eight in Central and South America, fifteen in Canada, thirteen in Africa, and seventeen in Asia/Australia (excluding India). The distribution for each country by scientific field was as follows. Mathematics: sixty-four in the United States and Canada, thirty-nine in India, thirty in Europe, four in Asia/Australia, and thirteen in Central and South America (giving a total of 152); physics: sixty-five in the United States and Canada, forty in India, forty-three in Europe, eighteen in Central and South America, seven in Asia/Australia, and four in Africa (a total of 177); chemistry: thirty-one in the United States and Canada, twenty-seven in India, seventeen in Europe, ten in Central and South America, two in Asia/Australia, and two in Africa (for a total of eighty-nine); biology: thirty-six in the United States and Canada, thirty in India, thirty-seven in Europe, seventeen in Central and South America, five in Africa, and four in Asia/Australia (making a total of 129). Given 28,856 as the total number of Jesuits in 1975, the Jesuits involved in science came to two percent, which is not a small number as the natural sciences are not the main focus of the Jesuits' work. This survey was carried out when the number of Jesuit scientists was at its highest, and the total number of Jesuits was still high (the highest number of Jesuits worldwide ever reached was 36,038 in 1965). However, the numbers in this survey are only approximate, since the survey was not complete, especially for science teachers in secondary schools.

The figures above contrast greatly with the present situation when considering how many Jesuits were active in scientific fields in the year 2005.[99] In that year, there were a total of 143—forty-seven in mathematics (including computer science), forty-two in physics, nineteen in chemistry, and thirty-five in biology. Clearly, the total number is much smaller than in 1975. The distribution is similar, but there has been an increase in the number of mathematicians vis-à-vis physicists because the number includes specialists in the new field of computer science. The geographic distribution is as follows: forty in Europe (including ten emeriti), of whom eleven are at the Vatican Observatory; sixty in the United States and Canada (including five emeriti); eighteen in India; fourteen in Asia/Australia; and eleven in Africa. As was the case in 1975, the two countries with the highest numbers are the United States with fifty-seven,

---

99   Data from the *Catalogs of the Provinces of the Society of Jesus* for the year 2005 (Archivum Romanum S.I., Rome). Students are not included.

including twenty-two mathematicians, and India with eighteen, nine of whom work in biology. Thus, between 1975 and 2005, there has been a notable diminution: in 2005, the number of Jesuits working in science is only a fourth of what it was in 1975. The largest decline has been in the United States, India, and Europe. However, numbers have increased in Africa and Asia/Australia, where numbers in 1975 were low. The average age of Jesuit scientists has also increased—in 2017, for instance, there were forty-two Jesuit scientists working at universities, but only thirteen are less than sixty years old.

The closure or transfer of observatories (about thirty-eight between 1960 and 1980) has inevitably affected Jesuit involvement in scientific work. Today, the only observatories with Jesuits on their staff are those of the Vatican and Manila. In the 1970s, many young Jesuits left the Society in Europe and the United States, and this also had an effect on those studying science. Those who remained in the Society and had begun their scientific careers in the 1970s were active until about 2000, and on retiring, they were often not replaced. Some young Jesuits were still studying science in 2005, but only a small number, so that while there will continue to be a few Jesuits engaged in science, they will be much fewer in number than in the 1970s and will not be present in as many countries.

These numbers must be compared with the total number of Jesuits over time. The overall number of Jesuits rose rapidly in the twentieth century: from about sixteen thousand in 1910 to a maximum of 36,038 in 1965, before declining steadily to 16,727 in 2015. The decline has been most dramatic in Europe: from eleven thousand in 1982 to 4,592 in 2015; in the United States, the number has fallen from 5,600 in 1982 to 2,607 in 2015.[100] In the years given for the two surveys considered above, 1975 and 2005, the total numbers of Jesuits dropped from 28,856 to 19,844 respectively, and the number of Jesuits in science dropped from 547 to 143. This represents a change from two percent to 0.7 percent of the total number of Jesuits. Thus, the overall number of Jesuits declined between 1975 and 2005 by thirty-one percent, while the number of Jesuit scientists in the same period has declined by seventy-four percent.

Although the decline in vocations has played an important role in the falling numbers of Jesuits involved in science, it does not fully explain what has happened, and other factors besides the demographic have to be considered to explain this decrease. Among them is the already mentioned influence of a change in apostolic priorities, which began with the Thirty-Second General Congregation in 1974 where a new emphasis was placed on the promotion of

---

100   Thomas Gaunt, "The Changing Jesuit Geography," February 1, 2011; http://nineteensixty-four.blogspot.com/2011/02/changing-jesuit-geography.html (accessed August 23, 2018).

faith and justice. Though unintentionally, this change has contributed to the overall decline. Today, Jesuit superiors no longer see the natural sciences as an urgent priority for Jesuits. The apologetic function that was so important in the Jesuit choice of scientific investigation at the beginning of the twentieth century, when the church was being accused of opposing science, is no longer such an important concern. New attitudes in the church, which originated from the Second Vatican Council, have changed the apologetic stance. Instead, there is a recognition of the mutual autonomy of science and religion and the need for a dialogue between them, in which Jesuits themselves have often become involved, as we have seen. Furthermore, science teaching in Jesuit universities and colleges is largely carried out by laymen, and Jesuit professors of science have now become very rare. The modern practice of scientific research has also changed in that it usually takes place in large institutions with considerable economic resources. Jesuits aspiring to work in cutting-edge scientific fields have little choice but to seek to join these institutions where there is usually strong competition for teaching and research posts.

Evidently, therefore, dedication to science is no longer seen as a priority among Jesuits. The recent *Yearbook of the Society of Jesus* (2017), for example, which lists the various activities in which the Jesuits are involved, makes no mention of scientific research. A recent trend among young Jesuits is for a greater interest in spirituality or pastoral work, which is often accompanied by a lack of interest in scientific research. Moreover, given the current tendency for men to join the Society at an older age, it is often more difficult for them to begin scientific studies unless they have been pursued beforehand, and even then, many who join the Jesuits with scientific or technological training are reluctant to continue in a scientific field and often opt to undertake direct pastoral work or studies in theology or spirituality.

14   Jesuit Scientists and Ignatian Spirituality

Having explored the long history of the post-restoration Jesuits' involvement in science, one question remains: What has motivated so many members of this order to engage in scientific work? Other religious orders have of course also produced noteworthy scientists: the Augustinian Gregor Mendel (1822–84), for instance, was a pioneer in the understanding of genetics; and, in more recent times, George Lemaitre (1894–1966), the father of the big-bang, was a diocesan priest. But the Jesuits are certainly unusual in that the Society has maintained an almost constant presence in scientific research from the very early years of its foundation up until its suppression in 1773, and this tradition was resumed

after the restoration in 1814, as we have seen. An indication of this continu-
ity can be seen in the fact that in 1906 Pope Pius X entrusted the Jesuits with
the direction of the Vatican Observatory only eighteen years after its foundation, and they have been in charge ever since. What, then, can explain this
phenomenon?[101]

Steven J. Harris was one of the first to try to find an answer to this question.[102]
He restricted himself to Jesuits of the seventeenth and eighteenth centuries,
but his ideas are equally relevant to modern times as well. In a detailed and
well-founded analysis, Harris puts what he calls "apostolic spirituality" as the
basis of the Jesuits' interest in science and sees this as the core of what he
calls "Jesuit ideology." First used by Rivka Feldhay, the term "Jesuit ideology"
refers to the Jesuit way of viewing knowledge as a path to salvation.[103] For
Harris, the main elements of the spirituality underlying the Jesuits' ideology
are the following: first, an emphasis on Christian *service*, which channels religious dedication outward, often into worldly activities not usually associated
with religious life; to this he adds an active engagement with the world, which
leads to a respect for *experience*, or a "testing against experience" and "proof
by trial," which provides standards of evaluation for ideas and projects. This
agrees with the methods used in modern science, where the empirical work
of observations and carrying out experiments are of key importance. Thus, for
Harris, this particular apostolic spirituality—the basic force behind the activity of every Jesuit—contains within itself the motivation for scientific work.
Another element is a special "esteem for learning," leading to what he calls the
"sanctification of learning," which is responsible for the enormous Jesuit educational endeavor that has characterized the order ever since its foundation,
with the creation of a global network of colleges and universities. According to
Harris, the sanctification of learning drove the Society to commit its members
to secular activities in the service of an ultimately spiritual goal: the salvation
of souls. He sees this as evident for the Jesuits of the sixteenth to eighteenth
centuries in three main areas: education, the European royal courts, and the
foreign missions—and the same also applies to the post-restoration Jesuits'
educational work and involvement in scientific research.

---

101   Daniel O'Connell, "Jesuit Men of Science," *Studies* 45 (1956): 307–18. Agustín Udías, "Jesuit
       Scientific Tradition and Ignatian Spirituality," *Lo Sguardo: Rivista di Filosofia* 10 (2012):
       207–19. George Coyne and Agustín Udías, "Spiritual Foundations for Jesuit Commitment
       to Science," *Conversations on Jesuit Higher Education* 47 (2015): 2–7.

102   Steven J. Harris, "Transposing Merton's Thesis: Apostolic Spirituality and the Establish-
       ment of the Jesuit Scientific Tradition," *Science in Context* 3 (1989): 29–65.

103   Rivka Feldhay, "Knowledge as Salvation in Jesuit Culture," *Science in Context* 1 (1987):
       195–213.

Historical factors have also played an important role. In the first place, there was the coincidence of the foundation of the Society of Jesus in 1540 and its choice of education in schools and universities with the beginning of modern science with the publication of Copernicus's work in 1543. When the Jesuits established their first schools shortly after their foundation, they did not have an established academic tradition that might have conditioned them, as was the case, for example, with the Dominicans and Franciscans. As a result, Jesuits could incorporate the new currents of modern science more easily into their teaching and thus respond to the social demands of their times, in particular with reference to the advances in mathematical and experimental sciences. Clavius, from his influential position as professor of mathematics in the Roman College, made a great effort to introduce the teaching of mathematical sciences into the first Jesuit colleges, and this was finally established as one of the norms for all Jesuit schools in the *Ratio studiorum*. Contrary to the predominant opinions in the traditional circles of his time, he insisted that mathematics was a true science, one that was essential for the study of natural philosophy. The introduction of mathematical sciences as an important element in the programs of Jesuit colleges was also an answer to the growing social importance that modern science was rapidly acquiring. Scheiner, who had a dispute with Galileo Galilei (1564–1642) about the discovery of sunspots, was aware of this social aspect and its apostolic importance. To Paul Guldin (1577–1643), professor of mathematics in the college of Graz, he wrote: "It is evident that mathematics is the net with which one can catch the elites and nobles and bring them to God's service."[104] Hence, all the way up until the suppression of 1773, Jesuit professors of mathematics saw their work as a natural part of their apostolic activity in the field of education.

After the Society's restoration in 1814, when the Jesuit scientific tradition was resumed, there was a new element in the role of Jesuits in the sciences, as we have seen, as it was partly motivated by an apologetic function in that they saw their work as a defense of the church against those who attacked it as an enemy of science. Jesuit scientists and the existence of their scientific institutions, such as their observatories, were a clear, tangible argument against these accusations while providing an example of the compatibility between science and Christian faith.

However, in addition to these historical factors, connected with the apostolic needs of educational work in colleges and universities, some more specific and intrinsic elements of Jesuit spirituality need to be considered. After

---

104   Steven J. Harris, "Les chaires de mathématiques," in *Les jésuites a la Renaissance*, ed. Luce Giard (Paris: Presses Universitaires de France, 1995), 239–62, here 253–54.

all, some elements of *apostolic spirituality* are to be found in other modern religious orders (i.e., those founded after the seventeenth century), some with an apostolic character similar to that of the Jesuits. But they have not had a similar engagement in the scientific world. How, then, is it possible to explain this peculiar tradition?[105]

The answer, as we saw in the introduction, can be found in what is known as "Ignatian spirituality," the foundations of which are based on the ascetic and mystical experiences of Saint Ignatius that can be found in his two great works: the *Spiritual Exercises* and the *Constitutions*.[106] The former outlines a month-long program of prayers and meditations, which is followed at the beginning and the end of Jesuit training and repeated annually for a period of eight days. It forms the core of Ignatian spirituality, and hence it is the Exercises that contain the motivation behind the Jesuits' unique devotion to scientific research. As mentioned earlier, the most important characteristic of Ignatian spirituality is that it is a mystique of *service*, understood as service to God through the apostolic ministry to men. This service must be understood as a service in the church.[107] The *Constitutions*, on the other hand, constitute the basic founding text of the order and define the aim of the Society of Jesus: "Not only to assist the salvation and perfection of its own members by divine grace, but with the same grace intensely to assist the salvation and perfection of their fellow men."[108]

In his writings, St. Ignatius often repeated the phrase that would later become the Jesuit motto: "All for the greater glory of God" (*Ad maiorem Dei gloriam*). Saint Paul urges the faithful to "do all for the glory of God" (1 Cor. 10:31), but St. Ignatius added the comparative "greater" to indicate that Jesuits should seek not just the glory but the *greater glory* of God. In the first meditation of the *Spiritual Exercises*, entitled the "Principle and Foundation," he exhorts that one should always act "desiring and choosing only what is 'most' conducive for us to the end for which we were created."[109] The term "more" (in Latin *magis*) indicates an important feature of this spirituality. In all times and circumstances,

---

105   Udías, *Searching*, 7–12.

106   Ignatius of Loyola, *Obras completas* (Madrid: Biblioteca de Autores Cristianos, 1991), 221–38, 432–646.

107   Jean-Claude Dhôtel, *La spiritualité ignatienne: Points de repère* (Paris: Vie Chrétienne, 1991); Rossano Zas Friz, "Espiritualidad ignaciana," in *Diccionario de espiritualidad ignaciana*, ed. José García de Castro et al. (Bilbao-Santander: Mensajero-Sal Terrae, 2007), 811–20.

108   Ignatius, *Obras* (Examen primero y general, chapter 1.3), 466.

109   Ignatius, *Obras*, 228–29.

Jesuits are inspired to seek this "more" or "greater" in their work. Even today, for many Jesuits, the realization of this goal is to be found in scientific work.

Another recurrent phrase in Ignatius's writings is "finding God in all things," which points to an awareness that God can be found in every person, place, and circumstance. Thus a Jesuit is asked to find God in whatever situation he may find himself. Accordingly, a Jesuit can maintain an attitude of prayer even in an active life. This is an important aspect in the last meditation of the *Spiritual Exercises*, the "Contemplation to Attain Love," where one is asked: "To look how God dwells in creatures, in the elements, in the plants, in the animals, in men [in me, myself] [...] and to consider how God works and labors for me in all things created on the face of the earth."[110] Thus all things, people, and circumstances are occasions for finding God, and scientific research is no exception. Moreover, with his emphasis on the humanity of Christ, St. Ignatius invites one to see that everything human, which includes science, can serve our encounter with God.

Jerónimo Nadal (1507–80), a companion of St. Ignatius, refers to what has been called the "Jesuit way" with the expression "contemplative in action,"[111] which implies a union between prayer and action; we can find God not only in the silence of prayer but also in the activity of work. Thus, there is no activity, no matter how profane, that cannot be transformed into prayer. Teaching mathematics or physics in a university, observing the light from a distant galaxy, or drawing a map of an unknown region are all perfectly compatible with a Jesuit vocation. In other religious groups, this may become a personal attitude, but for Jesuits it stems from the core of their spirituality. Angelo Secchi, the renowned astrophysicist, acknowledged this, saying: "The contemplation of God's works is one of the noblest works of the spirit; this is the principal aim of the study of nature."[112] Accordingly, Jesuits involve themselves in activities, among them scientific research, which others consider incompatible or at least unsuitable for a religious. Throughout their history, therefore, Jesuits can be found in active social work and as parish pastors, but also carrying out research in biology or astrophysics, or writing poetry, all motivated by the same spirituality of "seeking God in all things" for the greater glory of God.

Another pertinent aspect of Ignatian apostolic spirituality is its preference for situations and activities that may be called "frontier work": rather than devoting themselves to traditional pastoral activities inside the church, Jesuits try

---

110   Ignatius, *Obras*, 272–73.
111   Miguel Nicolau, "Contemplativo en la acción: Una frase famosa de Nadal," *Revista de espiritualidad ignaciana* (Center for Ignatian Spirituality) 25 (1977): 7–12.
112   Angelo Secchi, *Le soleil* (Paris: Gauthier-Villar, 1875), 6.

to get into contact with those who are outside the church in activities sometimes considered unconventional. Pope Benedict XVI (1927–, r.2005–13) recognized this when he said to the Jesuits: "As my Predecessors have said to you
on various occasions, the Church needs you, relies on you, and continues to
turn to you with trust, particularly to reach those physical and spiritual places
which others do not reach or have difficulty in reaching."[113] This explains how
Jesuits can sometimes be found praying in a Buddhist monastery or, conversely, carrying out physical research in a particle accelerator. Jesuits are always
searching for frontiers, for places and situations where the Christian message
is not yet known. It was this spirit, for example, that drove Matteo Ricci and his
companions in the seventeenth century to present themselves in the imperial
court of China as astronomers and mathematicians of the West, adopting the
dress and manners of Chinese scholars. In the nineteenth and the beginning
of the twentieth century, science was considered to be a field that was alien,
if not hostile, to religion, and Jesuits felt the need to show that there was no
incompatibility between science and faith, and that scientific work itself can
also be a vehicle to finding God. With this motive emerged the foundation of
a network of Jesuit observatories spread around the world. A modern Jesuit
scientist considers the scientific community to be his parish and overcomes by
his presence any barriers of estrangement or lack of understanding.[114]

15   Conclusion

As we have seen throughout this essay, from their restoration in 1814 to the
present, Jesuits have shown a continuous commitment to the natural sciences, a commitment unparalleled by any other religious order in the Catholic
Church, following the tradition of the order between the sixteenth and eighteenth centuries. Carried out largely as part of their educational commitments
in their global network of schools, colleges, and universities, this scientific
work has been marked by different characteristics over the years with a presence of Jesuits in science that has continued throughout the order's long history to the present day. The most remarkable feature of this commitment to
science is the creation of a worldwide network of seventy-four observatories
where research was carried out in astronomy, geomagnetism, seismology, and
meteorology, particularly in the period between 1820 and 1970, though such

---

113   *Congregación General 35 de la Compañía de Jesús* (Bilbao: Mensajero, 2008), 264.
114   Charlotte B. Harvey, "The Voyage of the Monte Carlo," *Boston College Magazine* (Fall 2000).

research also continues today at the Vatican Observatory. Other fields of particular Jesuit dedication are mathematics and biology at colleges and universities. Besides the pragmatic and apologetic motivation underlying this work, the Jesuits' involvement in science, as we have seen, needs to be understood in the wider context of Ignatian spirituality, the core of which lies in an emphasis on finding God in all things, the union of prayer and work, the search for what leads to the greater glory of God, and the preference for work "on the frontiers." This has often involved Jesuits in unconventional activities and situations, including scientific research. Jesuit scientists who have reflected on their work acknowledge this special affinity between the scientific vocation and their spirituality, although they are aware of the difficulties in combining this vocation with that of a Jesuit, of being at the same time priests and scientists. To conclude, the Jesuit scientific tradition, despite all the problems encountered during its long history, is still very much alive in modern times and serves as a special characteristic of the Catholic Church.

Bibliography

Altamore, Aldo and Sabino Maffeo, eds., *Angelo Secchi: L'avventura scientifica del Collegio Romano*. Foligno: Quarter, 2012.

Anduaga, Aitor, "Spanish Jesuits in the Philippines: Geophysical Research and Synergies between Science, Education and Trade, 1865–1898." *Annals of Science* 71, no. 4 (2013): 497–521.

"Approbantur statuta facultatum theologiae et philosophiae in collegiis Societatis Iesu erectarum," *Acta Romana S.I.* 7 (1934): 606–33.

Arrupe, Pedro, "El apostolado intelectual en la misión de la Compañía hoy," *Acta Romana S.I.* 16 (1973–76): 1015.

Bastero Monserrat, Juan Jesús, *Longino Navás, científico jesuita*. Zaragoza: Universidad de Zaragoza, 1989.

Berloty, Bonaventure, "L'Observatoire de Ksara (Liban)," *Ciel et Terre* 23 (1912): 103–8.

Birkenhauer, Henry, "Father Macelwane and the Jesuit Seismological Association," *Earthquakes Notes* 27 (1956): 12–13.

Bishop, George D., "Stephen Joseph Perry (1833–1889): Priest, Scientist, Educator." MSc diss., University of Manchester, 1977.

Bishop, George D., "Stephen Perry (1833–1889): Forgotten Jesuit Scientist and Educator," *Journal of the British Astronomical Association* 89 (1979): 473–84.

Bishop, George D., *Jesuit Pioneers of Modern Science and Mathematics*. Anand, Gujarat: Gujarat Sahitya Prakash, 2005.

Blum, Victor, "Sketch of the Life of James Bernard Macelwane, S.J.," *Earthquakes Notes* 27 (1956): 9–11.

Bole, P. V., "Rev. Fr. Hermenegild Santapau," *Taxon (International Association for Plant Taxonomy)* 19 (1970): 576–83.

Boudignon, Patrice, *Pierre Teilhard de Chardin, sa vie, son oeuvre, sa reflexión*. Paris: Éditions du Cerf, 2008.

Brannagan, David, "Earth, sky and prayer in harmony. Aspects of the interesting life of father Edward Pigot, S.J., B.A., M.B., B.C.H. (1858–1929). A Jesuit seismologist. Part I." *Earth Science History*, 29 (2010), 69–99.

Bulletin of the American Association of Jesuit Scientists (The Jesuit Science Bulletin), vols. 1–43 (1922–66).

Byerly, Perry and William Stauder, "James B. Macelwane, S.J., September 28, 1883–February 15, 1956," *Memoirs of the National Academy of Science* 31 (1958): 254–81.

Cabré, Ramón, "75 Años en la vanguardia de la sismología," *Revista geofísica* 29 (1988): 4–56.

Carlson, Per, "A Century of Cosmic Rays," *Physics Today* 65 (2012): 30–36.

Chinnici, Ileana, "An 'Italian observatory' in India: The History of the Calcutta Observatory," *Studies in History of Medicine and Science* 14 (1996): 1–21.

Chinnici, Ileana, "Il profilo scientifico e umano di Angelo Secchi." In *Angelo Secchi: L'avventura scientifica del Collegio Romano*, edited by Aldo Altamore and Sabino Maffeo, 43–64. Foligno: Quarter, 2012.

Coenraads, Robert R., "The San Calixto Observatory in La Paz, Bolivia: Eighty Years of Operation; Director Dr. Lawrence A. Drake, S.J.," *Journal and Proceedings, Royal Society of New South Wales* 126 (1993): 191–98.

Colin, Elie, *Les cyclones à Madagascar*. Tananarive: Imprimerie Officielle, 1905.

Combeau-Marc, Evelyne, "L'observatoire d'Ambohidempona (1888–1923): Pouvoir Jésuite et science colonial," *French Colonial History* 12 (2011): 103–21.

Congregación General 35 de la Compañía de Jesús. Bilbao: Mensajero, 2008.

Cortie, Aloysius L., "Angelo Secchi, S.J. (1818–1878)," in *Twelve Catholic Men of Science*, ed. Bertram Windle. London: Catholic Truth Society, 1923, 149–64.

Coyne, George and Agustín Udías, "Spiritual Foundations for Jesuit Commitment to Science," *Conversations on Jesuit Higher Education* 47 (2015): 2–7.

Cuénot, Claude, *Pierre Teilhard de Chardin: Les grands étapes de son évolution*. Paris: Librairie Plon, 1958.

Decaux, B., "Le R.P. Pierre Lejay (1898–1958)," *L'onde electrique* 284 (1959): 1–4.

Dechevrens, Marc, *Sur la variations de temperature observé dans les cyclones*. Shanghai: Observatoire de Zikawei, 1887.

Decreta Congregationis Generalis 28, 4. "De studiis," *Acta Romana S.I.* 9 (1938–40): 37–40.

Decreta Congregationis Generalis 31, "De labore et investigatione scientifica," *Acta Romana S.I.* 14 (1961–66): 958–60.

Devane, John F. and James P. McCaffrey, "Reverend Daniel Linehan (1904–1987)," *Bulletin of the Seismological Society of America* 78 (1988): 1618–20.

Dhôtel, Jean-Claude, *La spiritualité ignatienne: Points de repère.* Paris: Vie Chrétienne, 1991.

Dopp, H., "Johann Georg Hagen, S. J., 1847–1930," *Revue des Questions Scientifiques* 99 (1931), 5–37.

Drum, Walter M., *The Pioneer Forecasters of Hurricanes.* Washington, DC: Stormont and Jackson, 1905, 13–14.

D'Souza, Leo, "Jesuit Contribution to Environmental Protection," 2009; http://www .sjweb.info/sjs/networks/ecology/Indian%20Jesuits%20Contribution%20to%20 Environmental%20Protection.pdf (accessed August 23, 2018).

Due Rojo, Antonio, *Labor científica del R.P. Manuel Sánchez Navarro, S.J.* Granada: Observatorio de Cartuja, 1940.

Due Rojo, Antonio, "El R.P. Manuel M. Sánchez Navarro, S.J.," *Boletín Real Sociedad de Historia Natural* 39 (1941): 117–20.

Due Rojo, Antonio, "El cincuentenario del Observatorio de Cartuja," *Urania* 234 (1953): 67–81.

Dunn, Gordon E., "Tropical Cyclones." In *Compendium of Meteorology*, edited by Thomas F. Malone, 887–901. Boston: American Meteorological Society, 1951.

Ebel, John, "The Jesuit Science." *Company* (Summer 2009): 28–31.

Examen postulati Provinciae Belgicae de studio scientiarum physicarum a patribus eiusdem Provinciae institutum (1884). Archivum Romanum S.I. (Rome), *De scientiis, studia* 4–21 (1884–1921).

Feldhay, Rivka, "Knowledge as Salvation in Jesuit Culture," *Science in Context* 1 (1987): 195–213.

Fényi, Julius, "Die Sternwarte in Kalocsa, das Haynald observatorium," *Nachrichten der öster-ungarische Provinze* 3 (1906): 14–23.

Fernández Rodríguez, Jaime and Eduardo di Vita, *Presença dos jesuitas no mundo científico.* São Paulo: Edições Loyola, 2004.

Froc, Louis, *Atlas of the Tracks of 620 Typhoons.* Shanghai: Observatoire de Zikawei, 1920.

García Doncel, Manuel and Antoni Roca, *El Observatorio del Ebro: Un siglo de historia (1904–2004).* Roquetas: Observatorio del Ebro, 2007.

García-Herrera, Ricardo, Pedro Ribera, Emiliano Hernández, and Luis Gimeno, "Northwest Pacific Typhoons Documented by Philippine Jesuits, 1566–1900," *Journal of Geophysical Research* 112 (2007); doi:10.1029/2006JD007370.

Gaunt, Thomas, "The Changing Jesuit Geography," February 1, 2011; http://nine teensixty-four.blogspot.com/2011/02/changing-jesuit-geography.html (accessed August 23, 2018).

Gauthier, Henri, "Un maître en physique du globe (P. Marc Dechevrens)," *Études* 178 (1924): 272–91.

Gauthier, Henri, "Au service de tous et de chacun: Le pére Louis Froc (1859–1932)," *Études* 213 (1932): 273–86.

Gauvin, Jean-François, "La physique du globe tous azimuts: La carrière scientifique d'Ernesto Gherzi, S.J., 1910–1973," *Archivum historicum Societatis Iesu* 72 (2003): 45–80.

Geschwind, Carl-Henry, "Embracing Science and Research: Early Twentieth-Century Jesuits and Seismology in the United States," *Isis* 89 (1998): 27–49.

Gherzi, Ernesto, "Étude sur les microséismes," *Observatoire de Zikawei: Notes sismologiques* 5 (1924): 1–16.

Gherzi, Ernesto, "On the Constitution of Typhoons," *Bulletin of the American Meteorological Society* 19 (1939): 59–66.

Gutenberg, Beno and Charles F. Richter, *Seismicity of the Earth and Associated Phenomena*. Princeton: Princeton University Press, 1949.

Gutiérrez Lanza, Manuel, *Apuntes históricos acerca del Observatorio del Colegio de Belén, Habana*. Habana: Avisador Comercial, 1904.

Harris, Steven J., "Transposing Merton's Thesis: Apostolic Spirituality and the Establishment of the Jesuit Scientific Tradition," *Science in Context* 3 (1989): 29–65.

Harris, Steven J., "Les chaires de mathématiques," in *Les jésuites a la Renaissance*, ed. Luce Giard, 239–62. Paris: Presses Universitaires de France, 1995.

Harvey, Charlotte B., "The Voyage of the Monte Carlo," *Boston College Magazine* (Fall 2000).

Heck, Nicholas H., "The Jesuit Contribution to Seismology in the U.S.A.," *Thought* 19 (1944): 221–28.

Hennessey, James J., "Charles Deppermann S.J.: Philippine Scientist," *Philippine Studies* 5 (1957): 311–35.

Hennesey, James J., "The Manila Observatory," *Philippine Studies* 8 (1960): 99–120.

Héronnière, Édith de la, *Teilhard de Chardin, une mystique de la traversée*. Paris: Albin Michel, 2003.

Heyden, Francis J., *The Beginning and End of a Jesuit Observatory (1841–1972)*. Manila: Publication of Manila Observatory, 1975.

Ignacimuthu, Savarimuthu, "The Contributions of South Asian Jesuits to Environmental Work," *Journal of Jesuit Studies* 3, no. 4 (2016): 619–44.

Ignatius of Loyola, *Obras completas*. Madrid: Biblioteca de Autores Cristianos, 1991.

Imre, Mojzes, *A Kalocsai Haynald Obszervatórum Története*. Budapest: Orzágos Müszaki Információs, 1986.

Institutum Societatis Iesu. Vol. 2 Examen et constitutiones. Decreta congregationum generalium. Formulae congregationum. Florence: Typographia a. SS. Conceptione, 1893.

Jacquin, Robert, "Une polémique romaine sur l'hylomorphisme il-y-a un siècle," *Revue des sciences religieuses* 32 (1958): 372–77.

Janssen, Jean Baptiste, "Epistola de ministeriis nostris (22 June 1947)," *Acta Romana S.I.* 11 (1946–50): 299–336.

Janssen, Jean Baptiste, "Instructio de cosmologia in nostris scholasticatibus docenda," *Acta Romana S.I.* 13 (1960): 809–14.

King, Ursula, *Spirit of Fire: The Life and Vision of Teilhard de Chardin.* New York: Orbis, 1998.

Kozhamthadam, Job, "Jesuit Contribution to Science in India," *Vidyajyoti: Journal of Theological Reflection* 68 (2004): 892–916.

Kummer, Christian, "Forschung und Weltanschauung: die Rolle der deutschen Jesuiten in der Biologie," *Sammelblatt des Historischen Vereins Ingolstadt* 109 (2000): 105–20.

Lejay, Pierre, "Trois observatoires de mission," *Études* 182 (1925): 594–602.

Linehan, Daniel, "Jesuits in Seismology," *Earthquake Information Bulletin* 16 (1984): 156–65.

"List of Jesuit Institutions"; https://en.wikipedia.org/wiki/List_of_Jesuit_educational_institutions (accessed August, 2018).

L'Observatoire de Zi-ka-wei: Cinquante ans de travail scientifique. Paris: Imp. d'Art G. Boüan, 1930.

López de Santa Anna, Antonio, *Contribución a una biografía completa del P. Benito Viñes Martorell, S.I., célebre meteorólogo de las Antillas.* Santander: Hermanos Bedia, 1957.

Lustig, Abigail J., "Erich Wasmann, Ernst Haeckel and the Limits of Science," *Theory in Biosciences* 121 (2002): 252–59.

Lynch S.J., Joseph, *Watching Our Trembling Earth for 50 Years.* New York: Dodd, Mead and Co., 1970.

Macelwane, James B., "The Forerunner: The Jesuit Seismological Service," in *Jesuit Seismological Association, 1925–1950: Twenty-Fifth Anniversary Commemorative Volume,* ed. James B. Macelwane, 3–16. St. Louis: Saint Louis University, 1950.

Macelwane, James B., "The Jesuit Seismological Association," in *Jesuit Seismological Association, 1925–1950: Twenty-Fifth Anniversary Commemorative Volume,* ed. James B. Macelwane, 17–23. St. Louis: Saint Louis University, 1950.

Maffeo, Sabino, *In the service of nine popes. 100 years of the Vatican Observatory.* Vatican City: Vatican Observatory and Pontifical Academy of Sciences, 1991.

Maffeo, Sabino, *La Specola Vaticana. Nove Pappi, una missione.* Vatican City: Specola Vaticana, 2001.

Malone, Thomas F., ed., *Compendium of Meteorology* (Boston: American Meteorological Society, 1951). Especially the chapters by Gordon E. Dunn, "Tropical Cyclones," 887–901, and Herbert Riehl, "Aerology of Tropical Storms," 902–16.

Martelet, Gustave, *Teilhard de Chardin, prophète d'un Christ toujours plus grande*. Brussels: Lessius, 2005.

McCann, Charles, "Ethelbert Blatter, S.J." *Journal of the Bombay Natural History Society* 37 (1934): 465–73.

McKevitt, Gerald L., "Jesuits Schools in the USA, 1814–c.1970," in *The Cambridge Companion to the Jesuits*, ed. Thomas Worcester, 278–93. Cambridge: Cambridge University Press, 2008.

Milloux, Henri, "Notice sur la vie et les travaux de Pierre Lejay (1898–1958)," *Institut de France, Academie des Sciences* 7 (1961): 1–7.

Mitchell, Brian J. and Agustín Udías, "William Stauder (1922–2002)," *Seismological Research Letters* 74 (2003): 373.

Mohr, Paul, ed., *Geophysical Observatory of Addis Ababa University: 50th Anniversary Symposium Proceedings*. Addis Ababa: Addis Ababa University, 2007.

Mooney, Christopher F., *Teilhard de Chardin and the Mystery of Christ* (New York: Doubleday Image Book, 1968).

Murphy, Thomas, "The Rev. R. E. Ingram, S.J.," *Quarterly Journal of the Royal Astronomical Society* 9 (1968): 226–27.

"Necrological Session in Honor of Rev. Fr. Federico Faura, S.J.: Articles by Luis Rodes, Eduardo Fontseré and Eduardo Alcobé," *Memorias de la Real Academia de Ciencias y Artes de Barcelona* 22 (1930): 69–92.

Nicolau, Miguel, "Contemplativo en la acción: Una frase famosa de Nadal," *Revista de espiritualidad ignaciana* (Center for Ignatian Spirituality) 25 (1977): 7–12.

Observatorio de Cartuja, 1902–2002. Granada: Caja de Granada, 2002.

O'Connell, Daniel, "Jesuit Men of Science," *Studies* 45 (1956): 307–18.

Ott, Richard W., "Rev. Ramón Cabré Roigé, S.J. May 2, 1922–February 24, 1997," *Seismological Research Letters* 68 (1997): 901.

Palmés, F. M, J. Torra Almenara and J. M. Caballería, "Las 'cuestiones científicas' en las Facultades de Filosofía," *Estudios eclesiásticos* 55 (1935): 1–32.

Poisson, Charles, *Un cinquantenaire, l'Observatoire d'Ambohidempona, Tananarive, Madagascar, 1889–1939*. Paris: Editions Dillen, 1939.

Puig, Ignacio, "El P. José Algué, S.J.," *Ibérica* 833 (1930): 392–98.

Puig, Ignacio, *El Observatorio de San Miguel, República Argentina*. San Miguel, Observatorio de San Miguel, 1935.

Rabin, Sheila J., "Jesuit Science before 1773: A Historiographical Essay." In *Jesuit Historiography Online*, edited by Robert A. Maryks. https://referenceworks .brillonline.com/entries/jesuit-historiography-online/jesuit-science-before-1773-a -historiographical-essay-COM_196375 (accessed April 10, 2019).

Ramírez, Jesús Emilio, "An Experimental Investigation on the Nature and Origin of Microseisms at St. Louis, Missouri," *Bulletin of the Seismological Society of America* 30 (1940): 34–84; 139–78.

Ramírez, Jesús Emilio, *Historia del Instituto Geofísico al conmemorar sus 35 años (Universidad Javeriana) I. Antecedentes y fundación; II. Los años de la guerra 1944–1945; III. Consolidación del Instituto Geofísico.* Bogotá: Editora Guadalupe, 1977.

Ramos Guadalupe, Luis E., *Benito Viñes, S.J.: Estudio biográfico* (Havana: Editorial Academia, 1996).

Ramos Guadalupe, Luis E., *Father Benito Viñes: The 19th-Century Life and Contributions of a Cuban Hurricane Observer and Scientist,* trans. Oswaldo García. Boston: American Meteorological Society, 2014.

Repetti, William C., *The Manila Observatory.* Washington DC, 1948.

Riehl, Herbert, "Aerology of Tropical Storms." In *Compendium of Meteorology*, edited by Thomas F. Malone, 902–16. Boston: American Meteorological Society, 1951.

Romeiras, Francisco Malta, "The Journal *Brotéria* (1902–2002): Jesuit Science in the 20th Century," *Journal of History of Science and Technology* 6 (2012): 100–9.

Romeiras, Francisco Malta, "The Emergence of Molecular Genetics in Portugal: The Enterprise of Luís Archer," *Archivum historicum Societatis Iesu* 82 (2013): 501–12.

Romeiras, Francisco Malta, "A ciência da Companhia de Jesus nos séculos XIX e XX em Portugal," *Brotéria* 179 (2014): 429–54.

Saderra-Masó, Manuel, *Historia del Observatorio de Manila.* Manila: E. C. McCullough, 1915.

Sánchez Navarro-Neumann, Manuel M., "Os jesuitas e a sismologia," *Brotéria* 24 (1937): 145–51.

Schroth, Raymond A., *The American Jesuits: A History* (New York: New York University Press, 2007), 174–96.

Schumacher, John N., "One Hundred Years of Jesuit Scientists: The Manila Observatory," *Philippine Studies* 13 (1965): 258–86.

Secchi, Angelo, *Le soleil.* Paris: Gauthier-Villar, 1875.

Semaine d'étude sur le problème des microséismes, Scripta Varia 12. Vatican City: Pontificiae Academiae Scientiarum, 1952.

Speaight, Robert, *Teilhard de Chardin: A Biography.* London: Collins, 1967.

"Statuta Congregationis Generalis XXVII," *Acta Romana S.I.* 4 (1924): 59–61.

"Statuta Facultatum in Collegiis S.I. erectarum, Titulus V," *Acta Romana* 7 (1932–34): 630–31.

Tardi, Pierre, "Le R. P. Pierre Lejay. Notice Nécrologique," *Revue des Questions Scientifique* 130 (1959): 1–12.

Teilhard de Chardin, Pierre, *Oeuvres*, vols. 1–13. Paris: Éditions du Seuil, 1955–76.

Udías, Agustín, "Jesuits' contribution to meteorology," *Bulletin American Meteorological Society* 77 (1996): 2307–15.

Udías, Agustín, "Serving God and Science," *Astronomy and Geophysics* 42 (2001): 23–24.

Udías, Agustín, "Jesuits, Role in Geomagnetism." In D. Gubbins and E. Herrero-Berrera (eds.), *Encyclopaedia of Geomagnetism and Paleomagnetism*, 460–462. Dordrecht: Springer, 2007.

Udías, Agustín, *Searching the Heavens and the Earth: The History of Jesuit Observatories.* Dordrecht: Kluwert, 2003.

Udías, Agustín, "Lawrence Drake, S.J. (1931–2007)," *Seismological Research Letters* 78 (2007): 420.

Udías, Agustín, "Jesuits' Studies of Earthquakes and Seismological Stations," in *Geology and Religion: A History of Harmony and Hostility*, ed. Martina Kölbl-Ebert, 135–43. London: Geological Society S.P., 2009.

Udías, Agustín, "Jesuit Scientific Tradition and Ignatian Spirituality," *Lo Sguardo: Rivista di Filosofia* 10 (2012): 207–19.

Udías, Agustín, *Los jesuitas y la ciencia: Una tradición en la Iglesia* (Bilbao: Mensajero, 2014).

Udías, Agustín, *Jesuit Contribution to Science: A History*. Dordrecht: Springer, 2015.

Udías, Agustín, "The correspondence of Johann Georg Hagen, first Jesuit director of the Vatican Observatory," *Journal of Jesuit Studies* 3, no. 2 (2016): 259–278.

Udías, Agustín, "Jesuit Contribution to Science 1814–2000: A Historiographical Essay," *Jesuit Historiography Online*; http://referenceworks.brillonline.com/entries/jesuit-historiography-online/jesuit-contribution-to-science-18142000-a-historiographical-essay-COM_192552 (accessed August 2, 2018).

Udías, Agustín and William Stauder, "The Jesuit Contribution to Seismology," *Seismological Research Letters* 67 (1996): 10–19.

Vatican.va/holy_father_leo_xiii/encyclicals/documents/hf_l_xii_enc_04081879_aeternis_patris. No. 30. (Accessed February, 2017).

Vregille, Pierre de, "Les jésuites et l'étude du magnétism terrestre," *Études* 104 (1905): 492–511.

Vregille, Pierre de, "Les observatoires de la Compagnie de Jésus au debut du XXᵉ siècle," *Revue des questions scientifiques* 59 (1906): 10–72, 493–579.

Wideman, Charles J., *Erich Wasmann, S.J. 1859–1931*. San Francisco: Wasmann Biological Society, 1942.

Wulf, Theodor, *Die Faden-Elektrometer*. Berlin: Dümmlers, 1933.

Zas Friz, Rossano, "Espiritualidad ignaciana," in *Diccionario de espiritualidad ignaciana*, ed. José García de Castro et al., 811–20. Bilbao-Santander: Mensajero-Sal Terrae, 2007.

Printed in the United States
By Bookmasters